ROUTLEDGE LIBRAR'
POLITICAL GEOG

T0228049

Volume 11

THE GEOPOLITICS OF DOMINATION

THE GEOPOLITICS OF DOMINATION

GEOFFREY PARKER

LONDON AND NEW YORK

First published in 1988

This edition first published in 2015
by Routledge
2 Park Square, Milton Park, Abingdon, Oxon, OX14 4RN

and by Routledge
711 Third Avenue, New York, NY 10017

Routledge is an imprint of the Taylor & Francis Group, an informa business

British Library Cataloguing in Publication Data
A catalogue record for this book is available from the British Library

ISBN: 978-1-138-80830-0 (Set)
eISBN: 978-1-315-74725-5 (Set)
ISBN: 978-1-138-81332-8 (Volume 11)
eISBN: 978-1-315-74708-8 (Volume 11)
Pb ISBN: 978-1-138-81334-2 (Volume 11)

Publisher's Note
The publisher has gone to great lengths to ensure the quality of this reprint but
points out that some imperfections in the original copies may be apparent.

Disclaimer
The publisher has made every effort to trace copyright holders and would
welcome correspondence from those they have been unable to trace.

Printed and bound by CPI Group (UK) Ltd, Croydon, CR0 4YY

THE GEOPOLITICS OF DOMINATION

GEOFFREY PARKER

ROUTLEDGE
London and New York

First published in 1988 by
Routledge
11 New Fetter Lane, London EC4P 4EE

Published in the USA by
Routledge
in association with Routledge, Chapman & Hall, Inc.
29 West 35th Street, New York, NY 10001

Printed and bound in Great Britain by Mackays of Chatham Ltd, Kent

British Library Cataloguing in Publication Data

Parker, Geoffrey, *1933-*
    The geopolitics of domination.
    1. Geopolitics — Europe
    I. Title
    320.1′2′094   D104

ISBN 0-415-00483-7

Library of Congress Cataloging-in-Publication Data

ISBN 0-415-00483-7

CONTENTS

List of Tables
List of Figures
Preface

1. Bids for Supremacy: The Urge to Expansion
   and Domination   1

2. Empire and Domination in The Mediterranean   12

3. Empire and Domination North of the Alps   30

4. A Geopolitical Model of Dominance   64

5. From Ostrog to Empire: The Territorial
   Expansion of Russia   76

6. The Soviet Union: Socialist Commonwealth
   or New Imperial State?   102

7. The Silent Castle: A Case of Geopolitical
   Uniqueness?   133

8. Decline and Fall: The Retreat from Dominance   147

Epilogue: Gods, Men and Territory   160

Glossary   166

Index   173

TABLES AND FIGURES

TABLES

1. Geopolitical Characteristics of the Dominant
   State   66/67
2. The Soviet Union – Population and Population
   Change by Republic   115
3. Comparative Figures for Area, Population and
   Primary Production for the Soviet Union,
   France, West Germany and the United Kingdom   136
4. The Retreat from Dominance   150

FIGURES

2.1   The Geopolitical Centre of the Ottoman Empire   15
2.2   The Geopolitical Centre of the Spanish Empire   22
3.1   The Geopolitical Centre of the Austrian Empire   34
3.2   The Geopolitical Centre of France   37
3.3   The Geopolitical Centre of Germany   48
3.4   Sphere of German Territorial Dominance in 1942   54
4.1   Geopolitical Model of Dominance: Stage I   71
4.2   Geopolitical Model of Dominance: Stage 2   73
4.3   Geopolitical Model of Dominance: Stage 3   74
5.1   The Geopolitical Centre of Muscovite Russia   79
5.2   Geopolitical Characteristics of the Russian
      Empire West of the Urals in the early
      Twentieth Century   95
6.1   Western Frontier Changes after World War I   106
6.2   Western Frontiers and Sphere of Influence of
      the Soviet Union after World War II   107
6.3   Geopolitical Characteristics of the Soviet
      Union   120
7.1   The Soviet Union and the Western Ecumene   138

PREFACE

The attempts by certain states to achieve and retain a position of dominance over others has been a recurrent feature of world political history. One of the most important expressions of this has been territorial - the urge to secure control over large areas and so to establish empires of overwhelming size and power. In order to reach a deeper understanding of the international scene, both past and present, it is essential to address the question of the nature and causes of this urge to dominate. It has been associated with, among other things, political ambition, religion, militarism and greed. Perhaps above all it has been associated with the rise of the charismatic leader, the 'great man' who has taken his people in pursuit of some real or imagined destiny, and in so doing has set his imprint upon the age.

While these causes of the phenomenon of territorial dominance have deservedly been examined and debated with considerable vigour, specifically geographical aspects of the phenomenon have received far less attention. There has, it is true, been a nod in the direction of the 'geographical background', as though it were the stage set against which the action has taken place. However, little attempt has been made to examine holistically the geopolitical structures of those states which have attained positions of dominance, with a view to discovering what light such structures may be able to shed on the matter. The main purpose of this book is to attempt to rectify this omission.

While the object of study is the phenomenon of territorial dominance as a whole, this book will confine itself to an examination of its occurrence in the European and Mediterranean area - the western ecumene - since the Renaissance. It entails an examination of the geopolitical structures of those particular states which successively have succeeded in achieving a commanding position within it, and have then aspired to become the universal state. The particular states which have been considered as coming into this category are the Ottoman Empire, Spain, Austria, France and Germany. During the nineteenth century Great Britain achieved a position of over-whelming world dominance, but it was not a territorial one so far as Europe was concerned, so it has not been included in this study. The Russian Empire and the Soviet Union are examined with a view to discovering the incidence of dominant state characteristics which they display, and relating this to both Russian and Soviet international behaviour. The basic question being addressed is this: Is there a specific geo-politics of territorial dominance and, if so, to what extent does it help us to a better understanding of the whole phenomenon?

I acknowledge with gratitude the assistance given me during the writing of this book by the Royal Geographical Society and by the University of Birmingham. From the outset my wife, Brenda, has been an active participant and partner in the enterprise, and both her wide knowledge and her critical eye have made her willingly given advice and assist-ance invaluable.

Geoffrey Parker
Lichfield

ONE

BIDS FOR SUMPREMACY: THE URGE TO EXPANSION AND DOMINATION

'Life is a continuous sequence of dominations,' said the President of the General Assembly of the United Nations in his speech to the Second Session in 1947'.[1] The attempts by some states to achieve and retain positions of dominance have done much to shape the outlines of the world political map. The resistance of others to such domination, and their refusal to accept its consequences, has been a factor of equal importance. The United Nations, like the League of Nations before it, was originally established after a disastrous war in which certain great powers sought to achieve positions of dominance. In essence it represented the refusal of the states of the world to accept such domination, and an assertion, in an admittedly halting and uncertain manner, of their right to freedom and self-determination.

The opposing desires to control and to resist, to dominate and to be free from domination have acted dialectically upon one another throughout modern times. Geopolitically, they have been responsible for what Henrikson referred to as 'the neatly segmented, multicoloured world of the standard political map.'[2] Nowhere is the complexity of this more in evidence than in Europe and the Mediterranean, the western ecumene of the World-Island.[3] While the political map here is a palimpsest, retaining considerable evidences of older and largely superseded political forms, the standard units are now the 'neatly segmented' territorial states. Known, usually quite wrongly, as 'nation-states' or simply as 'nations', their preponderant cultural unity has been used as a basis for political unity and for the centralisation of control over a wide range of activities. As part of the process of control, qualities have been attributed to the state well in excess of those which it actually possesses. In this way popular support, sometimes of a highly emotional character, is gained for its continued existence and for the régime which controls it. This nationalism may then supply the emotional and intellectual foundations for a further enlargement of the state, and thus for the commencement or continuation of an expansionist process. Its initial object, whether consciously expressed or not, is to attain some ideal or idealised geographical territory in which the state will be in close conformity with the physical and human environment in which it exists. This entails the attainment of a territory in which there is a large measure of physical unity bounded by clearly defined 'natural' frontiers. Such conditions constitute the physical basis for what is envisaged as a more secure, prosperous and easily-governed state. Its ultimate justification has been expressed as the creation of that particular morphology which God, Nature or Reason was deemed to have ordained for it. Such a morphology may take the form of a homogeneous geographical area, a region of physical unity, an amorphous 'living space' or a shape which can be reduced to a geometrical figure, such as a triangle or a hexagon. Perhaps the most powerful factor of all underlying the search for the ideal morphology has been the myth of a national 'golden age' in which there was fulfilment and glory within a large and impressive homeland. Goblet saw this as 'super-imposing maps of dreamland empires' upon far bleaker contemporary realities,[4] a phenomenon which Kristof called 'the shift towards an idealised past ... when the fatherland and people were true to themselves'.[5] Whether it was space or time which was invoked, the central theme was what Ratzel called the 'Staatsidee,' a philosophical and moral conception of the mission and destiny of the state and the image of what it should become.[6] The nature of the national territory, and the natural environment occupied by the people then become woven into the fabric of the cultural heritage.

Such idealised morphologies can hardly be expected to fit together neatly like the pieces in some providential jigsaw puzzle, and the striving towards them has in practice usually brought states into conflict with one another. The possibility of armed confrontation has rarely deterred states from the pursuit of their territorial ambitions, especially when the prospect of material advantage has also beckoned. Force, or at least the threat of force, has been the method all too regularly employed in attempting to alter the frontiers of the state, and the stronger state is the one which is in the better position to attain its ambitions. While such territorial power politics has constituted the normal pattern of inter-state relationships, there has also been a widespread countervailing desire to see the continuation of Henrikson's 'neatly segmented multicoloured world'. The ambition of particular states to promote their own territorial advantage, and in so doing to ride roughshod over the others, has been checked by alliances of the threatened states. Thus the drive towards what is conceived as being the ideal state morphology has had to take into account the ability of other states, either alone or in combination, to set limits to its attainment. In the real world, frontiers have often been truce lines which have fallen well short of the ideal. When the balance of power has been seen to change, then eventually attempts have been made to redraw the frontiers so as to reflect the new situation. The durability of the multicoloured geopolitical surface at any time thus depends on the extent to which it accords with that 'geophysical and geosocial world'[7] which underlies it. If it does, then geopolitical stability will ensue; if it does not, even after a period of war, then further adjustment becomes necessary until the political surface conforms more closely to the other surfaces.

A major problem in bringing about adjustment is that the geopolitical surface tends to harden rapidly and then to assume the role of a given, a phenomenon as enduring as those natural features which are woven into its polychromatic patterns. The state exists in both space and time, and not only does it aspire towards the ideal morphology but also towards a rock-like permanence. The thought of decline and fall is altogether too painful to be contemplated with equanimity. Far too much has been invested in success by all classes of society for the retreat from greatness to be shrugged off as being part of a normal and inevitable process.

The geopolitical surface consequently possesses two particularly unstable characteristics. One is that since it is in the nature of the state to behave as though it were a permanent phenomenon and to put up considerable resistance to all changes which might be disadvantageous to it, pressures will tend to build up beneath the hardening political surface. These will then periodically erupt and in so doing cause severe damage to the demographic, economic and social surfaces constituting that 'geosocial world' which lies beneath it. The second unstable feature results from this disruption. Taking advantage of the disruption, certain states may then attempt to change the system definitively so as to promote their own particular advantage. Rather than working for peaceful change within the system, they choose to operate outside it and eventually to replace it with one constructed after their own image. In this way what may have started out as being an attempt to secure frontier rectifications and limited territorial gains may be transformed into the establishment of a regional hegemony. In some instances it may not stop there, and the expansionist state may then seek to establish a position of supremacy over a wider area. This constitutes domination, as it is understood in this book, and it is a phenomenon which has been frequently identified and variously explained. To Lord Acton it was a 'law of the modern world that power tends to expand indefinitely' and in so doing to transcend all barriers, abroad and at home, until itself met by superior force.[8] Braudel talked in vaguer terms of that 'hunger for the world' which was characteristic of expansionist states[9] and Martin Wight of the aspiration 'to become a universal empire'.[10]

What exactly is the cause of such 'hunger' on the part of certain states, this desire for expansion and universality? O'Sullivan has recently described it quite simply as being the consequence of that 'aggressive spirit' which has always been a fundamental driving force in world affairs. History, he asserts, 'is a striking record of the persistent desire of some people to lord it over others'.[11] But such an 'animus dominandi' relates not only to populations, but to territory as well. Since there is a historical distinction between 'regnum' and 'dominium,' the former implying rule over people and the latter over territory, the territorial imperative has always been a built-in feature of domination. The acquisition of territory, and ipso facto of everything on it, is the ultimate expression of the will to control. Unlike other more specific forms of control it constitutes what Sack has termed an 'open-ended' method of exercising control. Territoriality 'offers a means of asserting control without specifying in detail what is being controlled'.[12]

If it is to be more than simply a Vandal-like whirlwind conquest of brief duration and leaving little legacy, the exercise of such control must be based on something more substantial than either the 'aggressive spirit' or a 'hunger for the world'. To give it durability, it requires both a justification and an organisation to put it into practice. In Wight's words it must 'appeal to some design of international unity and solidarity'.[13] Justification has usually been provided by the state's invocation of noble and universal ideals, what Niebuhr called harnessing 'religious impulses and philosophies as instruments of its purposes'.[14] The object of organisation has been the creation of a centralised and uniform political structure through which control may most effectively be exercised. By placing its firm imprint on its conquests the expanding state aspires to a durability it would otherwise be unlikely to attain. An overall structure of this sort is what constitutes an empire.

The 'maius imperium' was originally the authority bestowed on its officials by the Senate of the Roman Republic for the discharge of certain specific commissions in its name. Its object was the furtherance of the 'res publica,' the general good, but in time the purpose of the authority bestowed tended to become less specific and to encompass broader objectives. Thus, well before the Roman Empire came into being, the imperium had come to possess a territorial sense, and the specific commission of the Senate had given place to the creation of a territorial state within the boundaries of which the exercise of authority was curbed by ever fewer restraints. From the time of Augustus the head of state was also given the title of imperator, the bearer of the authority of the imperium. Later, by assuming the additional title of dominus, he asserted complete authority over the territory of the empire. 'Regere imperio populos' implied absolute authority over all the people living inside the imperial frontiers. Rome thus assumed the position of hegemonial state of the Mediterranean, so achieving for the first time control of a region which had for long been an economic and cultural unit but, until then, never a political one. The impressive edifice was further extended to become the 'imperium orbis terrarum,' in effect the universal state of the western world. At its maximum territorial extent during the reign of the emperor Trajan it stretched from the north of Europe to the Sahara and from the Atlantic Ocean to Mesopotamia. While it centred on the Mediterranean coastlands, it embraced within its extended frontiers a variety of contrasting geographical environments. With the divine Imperator at its head, it was able to exercise virtually unlimited and 'open ended' control within these vast territories.

Long after its fall, the idea of the universal state continued to be a force in the western ecumene. It was kept alive both in the spiritual form of 'respublica Christiana,' deriving its authority from the Pope in Rome, and the political one of 'renovatio imperii Romanorum' - the reinstitution of the Roman Empire - embarked upon by Charlemagne's German successors. Along with the Byzantine Empire in the east, both sought to legitimise their existence by reference back to the imperium of the universal state of the ancient world. Following the failure to secure the longed-for unity, vernacular versions of imperium and Imperator were adopted by successor states possessing ever more tenuous links with Rome. During the nineteenth century the terms were again reactivated to indicate the ascendancy of the European powers over the rest of the World-Island. Such has been the impact of Europe's bid for world supremacy that the term 'imperialism' has since come to be closely identified with this particular phenomenon rather than with the far longer lineage of the attempts to establish a universal state within the western ecumene itself.

Lichtheim defined imperialism as 'the relationship of a hegemonial state to peoples or nations under its control'.[15] There are many different degrees of control, ranging from the Procrustean, seeking to impose absolute uniformity, to a loose 'primus inter pares' situation which may be relatively benign and tolerant by comparison. All, however, have the effect of removing rights and freedoms from the subject peoples and vesting them instead in the controlling power. The degree of control exercised will depend on the relationship of the process of expansion to the characteristics of the area in which it is taking place. The nature and strength of the expanding power will influence the extent to which it is able to sustain the process of expansion. The physical and human characteristics of the area will then either impede or stimulate the process. If process and territory sustain and stimulate one another then the expansion is likely to be successful and swift; if they do not it may be difficult and protracted. According to Modelski, the acquiescence and even the active support of the subject people is an essential pre-requisite for the maintenance of supremacy. 'Global power', he said, 'carried by a ruling nation cannot in the long run be supported solely by the people of that nation...In its re-lations with other peoples such power must satisfy them and give them an interest in the continuance and stability of the whole'.[16] Grenier, as quoted by Braudel, went even further than that by asserting that 'to be conquered a people must have acquiesced in its own defeat'.[17] By so doing, and at the same time accepting implicitly their own inferiority, the conquered people become the unwitting agents in the trans-formation of a hegemony into an empire.

Throughout the Europe-Mediterranean region, according to Stoianovich, there has been 'a force that was inimical to the very principle of universal monarchy, namely a strong tradition of opposition to territorial bigness, and to power without limits'.[18] This has led to considerable ambivalence on the matter of the optimum size of political structures, and of how all-embracing they should be. On the one hand there was the aspiration to the re-creation of an 'imperium orbis terrarum' in some form appropriate to the times. The Renaissance then gave birth to the idea of a united Europe founded upon the system of fairly equal and balanced terri-torial states. One of the more notable of such schemes was the Duc de Sully's 'grand dessein.' By the nineteenth century, with the sovereign states in the ascendant and the notion of a united Europe having been reduced, in Bismarck's phrase, to 'une fiction insoutenable' there still remained an aspiration towards an all-embracing structure which would curb the drift towards international anarchy. Originally taking the form of the 'Congress System', this soon became transformed into a 'concert' of the great powers designed to oversee the affairs of the continent. In the late nineteenth century, despite the growing rivalries of the great powers busily engaged in acquiring for themselves vast overseas empires, there remained an underlying desire for Europe to become something more than a fiction or a geographical ex-pression. Especially following the completion of what Lord Bryce called the 'World-Process',[19] and the new perception of the totality of the globe which this produced, there was a belief that the 'civilised' countries, the torchbearers of pro-gress in the world, should draw together in their own best interests, and especially in the task of illuminating that 'dark Egyptian night' which was believed to characterise most of the rest of the globe. Ironically, right on the eve of World War I, W.M. Ramsey had spoken of Europe's 'Imperial Peace' stretch-ing back through the Middle Ages to Rome, and looked forward to its re-establishment in a contemporary form in the not too distant future.[20]

Alongside this will to unity there remained widespread fears of its possible adverse consequences. To Acton it presented itself as the 'phantom of universal empire'. In his opinion 'the ancient belief in a supreme authority' could only be achieved 'at the expense of the equipoise of nations'.[21] As with the aspiration towards it, the fear of it also derived from Rome; not from the 'res publica' of Cicero but from the imperium of Caesar. Caesar's crossing of the Rubicon into Italia and his assumption of dictatorial powers was the prelude to Rome's occupation of the position of 'imperium orbis terrarum.' The fears of such 'caesarism' have focused on a succession of would-be European conquerors, but more sinisterly they invoke such awesome figures as Attila, Gengis Khan and Timur Lenk (Tamburlaine). The dread of the conquest of Europe from outside, and in particular from the inaccessible and mysterious depths of Asia, has always been a very real one. Too many Asiatic conquerors have been observed to begin their careers as popular heroes and to end them as bloodsoaked tyrants. European fears arising from the phenomenon of the world conqueror were clearly iterated by Marlowe when he made the fierce Tamburlaine proclaim:

> The God of War resigns his roume to me,
> Meaning to make me Generall of the world;
> Jove, viewing me in armes lookes pale and wan,
> Fearing my power should pull him from his throne.[22]

The principal European fears have thus not been of the idea of universal empire per se. At its best this idea has been identified with progress and prosperity and with that 'Imperial Peace' of which Ramsey spoke. The fear rather has been of the nature of the putative conquerors themselves and the likely adverse consequences of this for Europe. A widespread perception, usually supported by much empirical evidence, has been that expanding states rarely bode well for those unfortunate enough to lie in their paths. The central problem in the past, observed Louis Janz, has been that the growth of large states has been associated less with the creation of European unity than with such terms as expansion, annexation, conquest, Anschluss, penetration, occupation and protectorate. One might add with Marlowe, that it has been also associated with 'armes', 'war', 'griesly death' and, ultimately, 'the bankes of Styx'. All the terms used to describe political entities, said Janz, 'contain one common factor that has characterised the international relationships of European states during the past thousand years: the bid for supremacy, whether made or achieved by force'.[23]

The principal underlying cause of the European and World wars of modern times has been the refusal of the

7

international community to accept the legitimacy of such bids for supremacy. Any acceptance of the emergence of a new 'Imperator Caesar Augustus,' whether in the form of a Charles V, Louis XIV, Napoleon or Mussolini, has not been forthcoming. In the end the putative 'Generall of the World' has been removed from the scene and Acton's 'equipoise of nations' has been, at least for a time, restored. The resistance to domination has generally taken the form of ad hoc alliances of relatively less powerful states which have raised their banners in the name of freedom against tyranny. In such circumstances some small states have gained a power and recognition well beyond their relatively modest size, and national Davids have successfully beaten off the imperial Goliaths. By taking up the cudgels against the imposition of universalism by force, they have become the champions of the existence of a polychromatic Europe. The establishment of the sovereign state of Switzerland in the fastnesses of the Alps was an early example of the defiance of seemingly overwhelming power by a small band of medieval freedom-fighters. The Oath of the Rutli and the legend of William Tell are affirmations of the deep desire of the inhabitants of 'Das Haus der Freiheit' as Schiller called the Alps, to be free from the constraints of the universal state which enveloped them.

Between the disintegration of the limited universal states of the Middle Ages and the middle of the twentieth century, five states can be identified as having engaged in serious bids for supremacy over the western ecumene. These are the Ottoman Turks, Spain, Austria, France and Germany.[24] Although not one of them actually became an universal state, each of them attained a dominating position and mounted a powerful challenge to the continuation of the existing order. On each occasion this order was profoundly shaken and radical changes were brought about to the balance of power within it.

These five states had many similarities as historical phenomena. Consciously or unconsciously they took from Rome the concepts of 'imperium' and 'dominium' and used them to justify and to legitimise their territorial expansion. Vernacular forms of Caesar, Imperator, Dux and Dominus were used to express supreme power and such Roman symbolism as the eagle, the orb, the fasces and the triumphal arch became a part of the iconography. Their object was not to achieve the 'renovatio imperii Romanorum' in its medieval sense, but to use Roman concepts and symbols as instruments of their purposes. Their common lineage was both spatial and historical. Spatially, their principal area of operation was the western ecumene, while historically they sought to emulate the unique achievement of its only really successful universal state.

Yet in spirit these dominant states were very different both from Rome itself and from one another. Rome was always more the model than the mentor for subsequent imperialisms.[25] Their raisons d'être were a variety of Niebuhr's 'religious impulses and philosophies' and they included Christianity, Islam, Nature, Reason and racial superiority. In spatial terms not one of them actually attained the frontiers of the Roman Empire nor, on close scrutiny, did their morphologies bear very much resemblance to it. The territories which they succeeded at different times in conquering during their successive bids for supremacy stretched from Scandinavia to the Russian steppes and deeper into the Middle East than the Romans had ever successfully penetrated. However, the one common object of this heterogeneous band of conquerors was the achievement of dominance over the lands of the western ecumene.

'What is surprising', said Gould, 'is not the uniqueness of patterns of spatial organisation...but their extraordinary similarity... There are, perhaps, deep structures of human behaviour underlying these repetitive patterns'.[26] In order to test the validity of this in the area of state behaviour, an examination will be made of the geopolitical structures of the five states which have aspired to reach a dominant position. The main object of this will be to seek these 'repetitive patterns' and so to identify those spatial characteristics which have underlain the bids for supremacy.

NOTES AND REFERENCES

1. M. Wight, *Power Politics* (Royal Institute of International Affairs/Penguin, Harmondsworth, 1986), p.30.

2. A.K. Henrikson, 'The Geographical "Mental Maps" of American Foreign Policy Makers', *International Political Science Review: Politics and Geography*, vol. 1, no.4, 1980.

3. 'World-Island' was a term first used by Halford Mackinder to describe the single land-mass made up by Europe, Asia and Africa, See H.J. Mackinder, *Democratic Ideals and Reality; A Study in the Politics of Reconstruction* (Constable, London, 1919), pp. 81ff. The term 'ecumene' was coined by Derwent Whittlesey and he defined it as being 'the most populous region of a state, particularly that part most closely knit by communication lines'. See D. Whittlesey, *The Earth and the State: A Study in Political Geography* (Holt, New York, 1939), p. 597. The word is derived from the Greek *oikoumene*, meaning the whole of the inhabited world. Arnold Toynbee talked of 'the old world *Oikoumene*' as meaning the Mediterranean-Middle Eastern region which he saw as having possessed a fundamental historical unity. The use of 'western ecumene' in this book is an adaptation of this concept to the modern world scene, and is chosen in place of the clumsier 'Europe-Mediterranean region'. Roughly bounded by the Urals-Caspian line in the east and by the Sahara and its Middle-Eastern extension in the south, it stands in distinction

to the other 'ecumenes' of the World-Island: the 'eastern ecumene' centring on China and the Far East, and the 'southern ecumene' consisting mainly of the Indian sub-continent.

4. Y.M. Goblet, *Political Geography and the World Map* (George Philip, London, 1956), p.223.

5. L.K.D. Kristof, 'The Russian Image of Russia: An Applied Study in Geopolitical Methodology' in C.A. Fisher, *Essays in Political Geography* (Methuen, London, 1968).

6. F. Ratzel, *Politische Geographie* (Oldenburgh, Munich, 1897).

7. A.K. Henrikson, 'Geographical "Mental Maps"', p.497.

8. Lord Acton, *Lectures in Modern History* (Macmillan, London, 1906), p.51.

9. F. Braudel, *Capitalism and Material Life 1400-1800* (Collins Fontana, London, 1974), p.314.

10. M. Wight, *Power Politics*, p.37.

11. P.O'Sullivan, *Geopolitics* (Croom Helm, London, 1986), p.16.

12. D.R. Sack, *Conceptions of Space in Social Thought* (Macmillan, London, 1980), p.199.

13. M. Wight, *Power Politics*, p.37.

14. R. Niebuhr, *Nations and Empires* (Faber & Faber, London, 1959), p.127.

15. J. Lichtheim, *Imperialism* (Penguin, Harmondsworth, 1974).

16. G. Modelski, 'The Long Cycle of Global Politics and the Nation State', *Comparative Studies: Society and History*, no.20, 1978, p.234.

17. F. Braudel, *The Mediterranean and the Mediterranean World in the Age of Philip II* (Collins Fontana, London, 1972), p.1237.

18. T. Stoianovich, 'Russian Domination in the Balkans' in T. Hunczak, *Russian Imperialism from Ivan the Great to the Revolution* (Rutgers University Press, New Brunswick, New Jersey, 1974), p.205.

19. J. Bryce, 'The Relations of the Advanced and the Backward Races of Mankind, *Romanes Lecture* (Clarendon, Oxford, 1902).

20. W.M. Ramsey, 'The Imperial Peace', *Romanes Lecture* (Clarendon, Oxford, 1913).

21. Lord Acton, *Lectures in Modern History*, p. 50.

22. C. Marlowe, *Tamburlaine the Great*, Part 1, Act V, sc.i.

23. L. Janz, 'The Enlargement of the European Community', *European Community*, No. 1, 1973 (Office of the European Community, London, 1973).

24. A sixth state - Italy during the Fascist period - also sought to achieve a dominant position in the western ecumene. The principal object of its foreign policy was to become the leading power in the Mediterranean region. While

it possessed many of the geopolitical characteristics of a dominant state during the first half of the twentieth century, as a result of its near total lack of success in attaining its expansionist ambitions, it would have been quite unrealistic to have included it here.

25. As a result of both geography and ideology, Fascist Italy came closer to certain aspects of the Roman Empire than any of the five states examined. Fascism took its name from the *fasces*, the bundle of sticks with an axe, which was the symbol of Roman authority and of its *imperium*. Its national territory included *Italia*, that part of the Italian peninsula south of the Rubicon river which had been the centre of the Roman Empire, and the city of Rome was its national capital. Civic buildings in the Roman imperial style were constructed in Rome and elsewhere and grand processional avenues were laid out. The reactivation of the term *Mare Nostrum* as *Mare Nostro* was meant to show to the world that the Mediterranean was now considered to be an Italian sea, and that it was intended to embark upon the re-creation of the Roman Empire in a modern form. This ambition was forcefully expressed in cartographic terms in the gigantic mosaic maps of the growth of the Roman Empire which were erected in the Via dei Fori Imperiali. However, by the early 1940s it was only with considerable support from her ally, Germany, that Italy was able to maintain even a semblance of regional hegemony in the eastern Mediterranean, and this was soon to crumble in the face of Anglo-American maritime power.

26. D. Gregory, *Ideology, Science and Human Geography* (Hutchinson, London, 1978), p.104.

TWO

EMPIRE AND DOMINATION IN THE MEDITERRANEAN

During the second half of the fifteenth century the Ottoman Turks advanced westwards from Anatolia and during the following century went on to establish the largest state in the western ecumene and the principal challenge to the order then existing in the region. The Ottomans were former subjects of the Seljuk Turks who, along with the other Turkic peoples, had migrated into the Middle East from central Asia. In the eleventh century the Seljuk reached Asia Minor and there, abandoning their nomadic existence, they established the Sultanate of Rum on the borders of the Byzantine Empire. As the Byzantines contracted under this and other external pressures, Rum eventually extended its territory to cover most of the Anatolian plateau. The Ottomans, the people of Osman, first appear on the scene as military vassals of the Seljuk Turks who were brought from the east and given lands on the western frontiers. In these frontier regions with mixed populations they engaged in 'ghaza,' Islamic Holy War, against the Orthodox Christians of the Empire. Following the defeat of the Seljuks by the Tartars in 1307, Osman Ghazi proclaimed the independence of his people. At this time the continuing westward movement was bringing these Osmanlis ever closer to Propontis, the Sea of Marmara, around which lay the core region of the Byzantine Empire.[1] Osmanli territory centred on the lower Sakaria river, a major routeway from the Anatolian plateau to the coast, flowing into the Black Sea just to the east of the Byzantine city of Nicomedia. They established their capital at Yenisehir on a natural routeway connecting the Sakaria directly to the Sea of Marmara and a mere 100 kilometres from Constantinople itself. In 1326 Osman's son, Orkhan Ghazi, captured Bursa, the great Byzantine city just to the west of Yenisehir, and transferred his capital there. The capture of Nicaea and Nicomedia then followed, securing control over the southern shores of Marmara. By so doing the Osmanlis acquired a considerable slice of the Byzantine core region, an area possessing a large indigenous population and capable of generating considerable economic wealth, and one which was to prove an excellent base for further conquests.

In 1357 the Osmanlis crossed the Hellespontus (Dardanelles), captured Callipolis (Gallipoli), and thrust northwards into Rumeli, the Balkan mountain region. In 1361 Adrianople was captured and the capital was transferred there. Constantinople had been outflanked and, shorn of most of its imperial possessions, it was relegated virtually to the position of a city state. The Osmanlis then moved up the valley of the Maritsa into Macedonia and thence over the passes leading to the Vardar and Morava rivers. The Serbs, until then the dominant people of this mountainous region, were defeated at the battle of Kossovo Polje in 1389 and from then on the Osmanlis, or Ottomans, became overlords of the Balkans. Shortly after this the Mongols struck westwards and in 1402 defeated the Ottoman forces at Ankara in the Sakaria basin. As a result of this battle the eastern territories were for a time lost and the centre of gravity of Ottoman activities moved to the safety of the lands to the north of Marmara. Adrianople remained the capital until Constantinople at last fell to Sultan Mohammed II, 'The Conqueror', in 1453 and the Ottoman capital was then transferred definitively to the former Byzantine capital. The Ottomans followed this up by the occupation of the Byzantine possessions in Greece and moved northwards into the Danube basin. The occupation of the Pannonian plain followed the defeat of the Magyars at the Battle of Mohacs in 1526. The large and impressive kingdom of Hungary was incorporated into the empire and three years later Vienna, the Habsburg capital and seat of the Holy Roman Emperors was besieged but not captured.

During the first quarter of the sixteenth century the Ottoman armies turned southwards and overran Syria, Egypt and northern Mesopotamia. In 1517 the Sultan Selim had been proclaimed Caliph, the successor to the prophet Mohammed, thus assuming the role of leader and protector of Islam. Ottoman rule was then extended over the whole of Mesopotamia and around the southern shores of the Mediterranean as far as the Maghreb. Here their sea power, until then relatively weak and ineffective, was greatly enhanced by the addition of the Barbary Corsairs, who were from then on to make a major contribution to the strength of the Ottoman fleet. At the Battle of Preveza in 1538 this fleet defeated the combined naval forces of Spain, Genoa and Venice, and from that time on the Ottomans became unchallenged masters of the eastern Mediterranean, posing a considerable threat to Italy, the heart of the European world. The late sixteenth century saw the Ottoman Empire at the height of its power, covering a territory of some five million square kilometres and having a population of 28 millions.[2] It dominated the Middle East, eastern Europe and most of the Mediterranean Sea, and under

Sulaiman I, 'The Magnificent', became a contender for the role of universal state of the western ecumene.

The process of the transformation of a small and impoverished tribal state into so formidable an empire can be traced back to the geographical environment in which it began. The Osmanlis were a frontier people on the edge of an aggressively expanding Islam and they formed the principal buffer between the Seljuk and the Byzantine spheres. These troubled Anatolian frontiers were, as Braudel put it, 'a rendezvous for adventurers and fanatics' and this generated 'unparalleled mystical enthusiasm'. 'These beginnings', he said, 'gave the Ottoman state its style'.[3] The actual frontiers were indeterminate, the political situation fluid and the land unable to sustain more than a bare existence. In this unpromising environment these primitive and violent nomads were nevertheless able to make good use of their speed, mobility and fighting qualities in the first instance to wage aggressive warfare in the name of Islam and then to secure a dominant position over their coreligionists. While they affected a disdain for wealth and material goods, the gradual acquisition of richer lands, as they filtered down from the Anatolian plateau to the coastal plains, enabled them to conduct their 'ghaza' more successfully and, eventually, on a far larger scale. As they pushed westwards into the heartlands of the shrinking Byzantine Empire they crossed an important junction between the two contrasting geographical environments.[4] The Anatolian plateau from which they came is mostly over 1000 metres in height, has an annual rainfall at its centre of under 250 millimetres and summer temperatures averaging over 25°C. These harsh conditions support a steppe and semi-desert vegetation of short grass and scrub. The coastal plain around Marmara presents a considerable contrast to this. It is between 50 and 100 kilometres wide and has a Mediterranean-type climate with a rainfall of over 500 millimetres, most of it falling in winter, and a long growing season. While the Anatolian plateau had historically supported only relatively poor pastoral societies, the coastlands in contrast constituted a part of the Graeco-Roman world with its flourishing commerce, large cities and intensive agriculture. The Ottomans thus moved into a rich and unfamiliar geographical environment into which they came as both conquerors and proselytisers for their clear and simple faith born of the desert. Once there, they assumed the role of a theocratic warrior aristocracy, ruling as once they had themselves been ruled. Rapidly appreciating the value of the coastal plains, they soon acquired the whole of the Marmara politico-economic core region and this became their base for the conquest of the Balkans.[5] Rumeli is in most ways a physical and cultural extension of the Marmara region, bound to it and to the northern Aegean by the valleys of the Maritsa, Struma and Vardar rivers, which penetrate deep into

Figure 2.1: The Geopolitical Centre of the Ottoman Empire

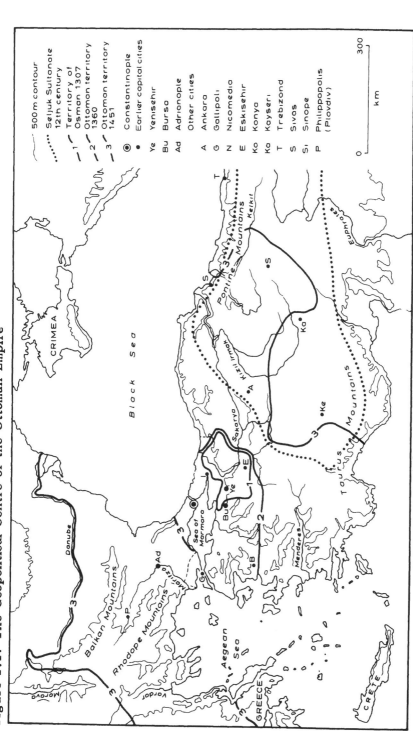

its mountainous centre. There was a gradual transition from the congenial environment of the coastlands to the harsher conditions of central Europe. The power centring on Marmara had historically been able to dominate both Anatolia and the Balkans. While Anatolia was the historic heartland for the Ottoman Turks, the Balkans came to be one of their richest and most prized possessions. It was adjacent to the capital and easily accessible through the river valleys leading up into it. Its well-watered plains with their lengthy growing season made it, in Braudel's phrase, 'the granary of Turkey'.[6] It had a large indigenous population much of which was initially well disposed to the Ottoman conquest[7] and the 'pax Turcica' which ensued. It was to become an important source both of revenue and of servants for the Ottoman state. It was also well away from the historic routeway east of the Caspian through which the fierce nomads had streamed out of the centre of Asia wreaking havoc in the Middle East. It was just such an occurrence, the return of the Tartars, which had been the initial impetus for the movement westwards. For the Ottomans 'geography and history alike', as Newbigin has put it, 'pointed to the Balkan peninsula'.[8] Together with the adjacent coasts, it was to become the Ottoman Empire's largest single source of wealth of a magnitude greater than that available to any other Turkic people. This was used to help pay for the conquests in the east which were initially motivated by the assumption of the Caliphate by the Sultan and the religious responsibilities which this office brought with it.

The despotic government of the Empire was inherited from the early Osmanlis, but, following the conquests, it was perpetuated and even strengthened by the device of enlisting the subject peoples directly into the service of the state. Of particular importance was the new military corps of Yeni Ceri (Janissaries) which was drawn from children of Christian subjects of the Sultan, and for these the Balkans was the most important recruiting ground. The people of this region also provided recruits for the higher ranks of the administration and the diplomatic corps. The autocracy centring on the Sultan arose from the close alliance of the temporal and religious establishments. This was also inherent in the early Osmanli state, but it was reinforced by the theocratic structures adopted in the wake of the assumption of the Caliphate. Islam implies submission, and there was initially little resistance by the Turks or by their Islamic subjects to the arbitrary exercise of power by their rulers.

The territorial expansion of the Ottoman Empire took place in four principal stages. In the first of these they acquired the Marmara region, and by so doing ended their Anatolian isolation and established themselves in the rich and cosmopolitan core of the Byzantine Empire[9] (Figure 2.1). The second stage consisted of the twin movements out from this imperial core westwards into the Balkans and eastwards back into Anatolia. By securing the latter the Ottomans controlled the lands of their former masters, the Seljuks, and became the successor state to the Sultanate of Rum. Thus by the middle of the fourteenth century the Ottomans had acquired territories virtually coextensive with those of the Byzantine Empire at the time of the first appearance of the Seljuks some four centuries earlier. The next stage consisted of the advance into the middle Danube basin and the securing of control over the eastern Mediterranean, this latter making them the successors to the Mameluke Sultanate. In the fourth stage, which can be taken as including the assumption of the Caliphate itself, the Ottoman Empire became also the successor state to the Arab Empire. It actually surpassed its great predecessor which had never succeeded in subduing the Byzantine Empire. In their move towards acquiring complete mastery of the Mediterranean, the Turks drew on the talents of the indigenous population. Such were their initial successes in this new environment that, by the middle of the sixteenth century, the Mediterranean Sea began to take on the appearance of an Ottoman 'Mare Internam'. While Italy retained its position as the heart of Christendom, not only in the cultural and religious, but also in the economic sense, it gained a new and unwelcome position as the embattled frontier in the conflict between the Cross and the Crescent. The seas around the peninsula became the focus of the Ottoman endeavour to supplant Christianity by Islam. Following the fall of Constantinople, the Sultan Mohammed II saw himself as heir to Byzantium. Ambitious to establish a 'universal empire' with his new capital at its centre, he accepted learned assurances that, as a result of the conquest, he had become legally 'Emperor of the Romans'.[10] His successor Sulaiman looked further west to Rome itself, the legendary 'red apple', and desired its possession to confirm his empire's position as universal state of the west. In this he was not successful and the powers of Christendom were able to keep hold over one corner of the Mediterranean until such time as they once more became strong enough to mount a counter-offensive.

The basic geographical pattern of Ottoman expansion was a concentric one out from the Marmara core region. Nonconformity of the actual morphology of the Empire with this model is attributable to varying speeds of conquest in different directions.[11] The conquests to the south and east were relatively swift and easy while those to the north and west were brought to a halt by the increasing physical and human difficulties encountered. The southern advance had been into the semi-desert regions around the Mediterranean, which were relatively lightly populated by peoples having a material culture and military strength inferior to those of the Ottomans themselves. Since the time of their initial conquest by the Arabs they had remained basically Arabic in language and culture and Islamic in religion, and in this context the

Ottoman conquests could be presented as being a reunification within a new theocratic state. In contrast, the western advance ground to a halt in the face of difficult and unfamiliar physical conditions and peoples who had become increasingly hostile to Ottoman rule. 'In the Balkans', said Braudel, 'geography created multiple autonomous zones'[12] and, especially in the mountainous fastnesses of the centre, the Turks found it almost impossible to maintain firm control. Further north, the Pannonian plain, outlier of the Asiatic steppes, was physically easier, but it was flanked to the north and west by the alien environment of central Europe. Here they came into contact and confrontation with the Holy Roman Empire, which had for centuries been engaged in active proselytising for Catholicism and, in so doing, advancing its frontiers eastwards. The Germans proved to be a more formidable foe than had been either the Byzantine Greeks or the Balkan Slavs, and they brought the Ottoman advance to a halt in front of Vienna.[13] For a century and a half Germans from the middle Danube and Turks from Anatolia confronted one another along that extended military frontier which Metternich in the nineteenth century still saw as being the real boundary between Europe and Asia.

Further south, in the Mediterranean region, their westwards advance took the Turks into an environment which was physcially even more alien to a people coming from the arid interior of the Middle East. Their success came to depend on active support from the indigenous inhabitants of the region, in particular the Greeks and the Barbary Corsairs. Their considerable maritime expertise was to make the Ottoman fleet the most feared in the Mediterranean. However, the victory at Preveza was the high point of their naval success and, despite their most strenuous efforts, they subsequently failed to take Malta. This brought to a halt their attempt to secure control over the western basin of the Mediterranean, and their defeat at Lepanto in 1571 by the fleet of the Holy League of Spain, Venice and the Pope effectively ended their bid for maritime supremacy. The countervailing naval power of the Europeans had proved too strong for them.

The formidable power of the Ottomans, so overwhelming when deployed against the Byzantines and the heirs to the Mamelukes, became less effective in unfamiliar environments. 'The Turkish conquests placed the lowlands in serfdom', said Braudel, but in the mountains their grip became 'visibly weaker'.[14] Likewise their grip was less certain at sea than on land and their naval success in the western Mediterranean was very limited. The actual physical power deployed against them at the two sieges of Vienna and Malta was considerably less than they themselves had at their disposal, and even after defeat they still had sufficient resources to raise new armies and fleets. However, geography, their ally east of the Aegean, became their enemy on the Danube and helped bring about their downfall in the Sicilian Narrows. It enabled the politically divided and numerically inferior Christian powers to frustrate the ambition of Sulaiman, 'Hero of Creation, Champion of the Earth and Time... Sultan of the Mediterranean and the Black Sea' to rule over the first universal state in the western ecumene since Rome. Although a formidable and impressive structure, the 'pax Turcica' failed to achieve the geographical completeness of the 'pax Romana', and the Mediterranean Sea remained divided.

The Iberian peninsula lies at the opposite end of the Mediterranean from Anatolia. It was here that the power arose which was to be the principal adversary of the Ottomans and their chief contestant for domination over the west. In both area and latitudinal range Iberia and Anatolia are very similar, but there is a significant difference in their relationship to adjacent areas. While Anatolia merges at its eastern extremity into the latitudinally oriented mountains of the Middle East, the mountains of the western Mediterranean, likewise oriented latitudinally, have had the effect of isolating Iberia from its neighbours. Within its total area of 600,000 square kilometres, Iberia possesses considerable physical variety, but the most significant division is that between the interior of the peninsula and its coastlands. In the centre is the Meseta, a large plateau which ranges in height from 300 to 1000 metres, and which is bounded and traversed by high mountain ranges also oriented from west to east. In the north are the Cantabrian mountains, in the south the Sierra Morena and in the centre the Sierra de Guadarrama and Sierra de Gredos. The Meseta is deeply dissected by a series of river valleys which, together with their tributaries, cut the plateau into distinct basins. The largest are those of the rivers Douro, Tagus, Guadiana and Guadalquivir which flow westwards into the Atlantic. Consequently the Meseta is oriented towards the west and here the coastal plains are largest, while the highest land and the principal watershed lies well to the east. The only really important exception to this orientation is the River Ebro which flows south-eastwards from the Cantabrian mountains into the Mediterranean and which forms a large basin between the central Sierras and the Pyrenees. The climate of the Meseta is continental, with hot summers, large temperature ranges and an annual rainfall which makes it one of the driest places in the whole of Europe. In contrast, the climate in the east and south of the coastlands is Mediterranean in type, while in the west, facing the Atlantic, there are maritime conditions with heavier rainfall.

In the early years of the eighth century most of the Iberian peninsula was conquered by the Arabs. They brought with them the religion of Islam and the remnants of the Christians were relegated to the extreme northern coastlands beyond the Cantabrian mountains. The conquered lands became Al-Andalus, and the Arabs entrenched themselves most strongly in the south with their principal centre at Cordoba, in the rich and productive Guadalquivir basin. The Christian counter-offensive began with Charlemagne who established his 'Marca Hispanica' to the south of the Pyrenees as the forward frontier of Christendom against the forces of Islam. It was from here, together with the small Christian states in the Cantabrian mountains, that the Reconquest – Reconquista – began. By the thirteenth century the multiplicity of small states had been consolidated into three larger ones, these being Portugal in the west, Aragon in the east and Castile in the centre. The two peripheral states came to play a more limited role in the later campaigns, their energies being increasingly diverted into maritime activity in the Atlantic and Mediterranean respectively. This was to make them by the fifteenth century into important commercial states having substantial interests outside the Iberian peninsula itself. In contrast, the paramount interest of landlocked Castile remained the Reconquista, and the state was organised to achieve this objective. From the initial construction of the fortified towers on the southern flanks of the Cantabrians, Castile had been a state founded on a military principle. It was ruled by a warrior aristocracy in close association with the Church, and while its resources were directed initially towards the creation of a 'front line of vigilant cities',[15] this led on to Crusade and conquest on the southern frontier. Old Castile, located mainly on the Meseta north of the central Sierras, consisted largely of latifundia given over to pastoral farming, and supplying the material resources for the incessant campaigns. Steadily the frontier moved southwards and in 1085 Toledo, the ancient Visigothic capital, was captured and the Castilian capital was transferred there from Burgos. It was from Toledo that the last stage of the Reconquista was directed, and Arab power was broken with the fall of Cordoba in 1236. The Arabs were not completely removed from Andalusia, and the vestigial Kingdom of Granada continued to exist in the southern mountains beyond the Guadalquivir for another two and a half centuries.

The political unification of the Iberian peninsula then took place in three stages. In 1479 the Crowns of Castile and Aragon were united through the marriage of Queen Isabella and King Ferdinand, and 'Los reyes católicos' then went on in 1492 to complete the Reconquista with the annexation of Granada and the subsequent expulsions of Jews and Moors. Finally, in 1560 Philip II of Spain became King of Portugal, so uniting the whole Iberian peninsula under Castilian dominance. In the previous century it was the union with Aragon which had ended the continental isolation of Castile and involved her in Mediterranean affairs. Already in the thirteenth century Aragon had acquired a large maritime commercial empire which included the Balearics, Sardinia and Sicily.

The unified Spanish state then went on to acquire territories on the mainland of Italy itself, including the Kingdom of Naples and the Duchy of Milan, and Spanish garrisons were also stationed in Rome and Florence. By the middle of the sixteenth century Spain had thus become the hegemonial power of the Italian peninsula and, through her control of the islands and her garrisons on the north African coasts, overlord of the whole of the western basin of the Mediterranean. As has been seen, the Ottoman Empire was the dominant power in the eastern basin of the Mediterranean and posed a considerable threat to the vulnerable heart of Christendom, the peninsula of Italy. With the Reconquista, for so long the exclusive preoccupation of the Castilian state, finally at an end, a new area appeared on the horizon towards which the martial temper of the Castilian warrior aristocracy could appropriately be directed. Fresh from their glorious but easy triumph over Granada, their Most Catholic Majesties then switched the forces of the Reconquista into the very heart of what was then Christendom, 'the threatened citadel of Italy'.[16] It was the King of Spain who provided the backbone for the Holy Alliance of Christian powers which halted the Ottomans at Malta and defeated them at Lepanto.

The accession of Charles V to the thrones of both the Holy Roman Empire and Spain had the effect of drawing Spain into the affairs of northern Europe at just the time when the authority of Pope and Emperor was coming under severe threat north of the Alps. The response of Spain to the challenge of the Reformation was, in effect, to carry the spirit of the Reconquista across the Alps and to direct it against those heretics who threatened the unity of Christendom and seemed to be bringing about the dilution of the purity of Christianity. Spain, master of Italy and the western Mediterranean, switched her attention to the enemy within.

The largest and richest of the Spanish acquisitions in northern Europe was the Netherlands, a hereditary possession of the House of Habsburg, and this was used as the principal base for interference in the affairs of northern Europe. Its land link with the Mediterranean was the so-called 'Spanish road' stretching from northern Italy through the Alps to the lower Rhinelands.[17] This was kept open by the military power of Spain, in particular the much-feared tercios, the greatest infantry of the age. Like most of northern Europe, the Netherlands had been influenced by the Reformation, and in 1579 the United Provinces declared themselves independent. The Spanish reply was to reassert their dominance by force of arms.

In the same year as the subjugation of Granada, Christopher Columbus sailed westwards from Palos in Andalusia and discovered the existence of the New World on the other side of the Atlantic. This he claimed for 'Los reyes católicos' and within a decade the Spaniards had become

Figure 2.2: The Geopolitical Centre of the Spanish Empire

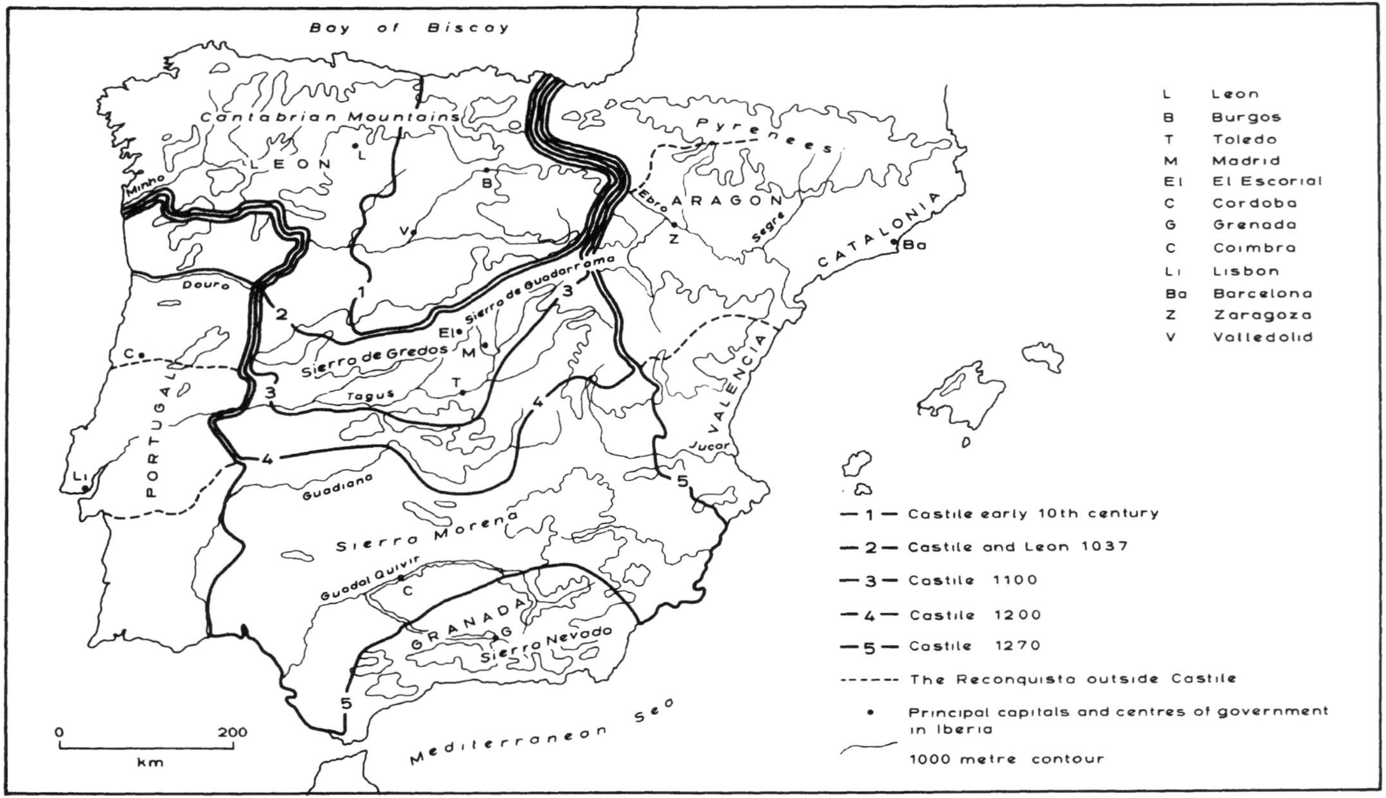

L Leon
B Burgos
T Toledo
M Madrid
El El Escorial
C Cordoba
G Granada
C Coimbra
Li Lisbon
Ba Barcelona
Z Zaragoza
V Valledolid

—1— Castile early 10th century
—2— Castile and Leon 1037
—3— Castile 1100
—4— Castile 1200
—5— Castile 1270
------- The Reconqusta outside Castile
• Principal capitals and centres of government in Iberia
——— 1000 metre contour

Bay of Biscay

Cantabrian Mountains

Pyrenees

ARAGON

CATALONIA

Segre

Ebro

VALENCIA

Jucar

LEON

Douro

Minho

PORTUGAL

Sierra de Guadarrama

El Sierra de Gredos

Tagus

Sierra de Gredos

Guadiana

Sierra Morena

Guadalquivir

GRANADA

Sierra Nevada

Mediterranean Sea

0 200 km

deeply involved with their new possessions in the name of the Pope and for the greater glory of Christendom. By the middle of the sixteenth century, the Spanish Conquistadores had claimed immense territories for the Spanish crown in both south and central America. Thus at just the time when she became the hegemonial power of the western Mediterranean and the Rhinelands, Spain also became Europe's first global power and the forerunner of maritime imperialism. Yet, despite this impressive imperial edifice, the Spaniards failed to achieve the kind of dominance over the affairs of Christendom which the Ottomans exercised over those of Islam. The Spanish period of hegemony also lasted for a considerably shorter time than that of the Ottomans; by the early seventeenth century it had been brought to an end. Both its meteoric success and its lack of completeness and durability will be examined in the geopolitical context.

The historic core of the Spanish state lay on the northern Meseta, and it was from here that political control was exercised first over Iberia and then over the growing overseas possessions[18] (Figure 2.2). The nature of the state which emerged on this remote and isolated fringe of Europe was militaristic, aristocratic, fanatically Catholic and based on a latifundist rural economy. Here on the frontier of Christendom and Islam the Castilians refined their austere faith in their battles against the Infidel. As Wittfogel has observed, it was its very success in war and conquest which strengthened absolutism and enhanced the growth of royal authority.[19] It also increased the preoccupation with Catholicism and with the purity of the faith. This developed into a conviction that purity could only be achieved through purity of the blood, 'limpieza de sangre', and an assertion that only the Castilians were the original Christians, the 'Cristianos viejos'. In turn this led to an overpowering will to dominate, to impose the hard-won faith on all other peoples, first in the form of domination over the Iberian peninsula and later over the western Mediterranean, northern Europe and the New World. Its origins being crusading and militaristic, its aristocracy had long developed a contempt for the easy and sedentary life. This percolated down the social strata to become 'hidalguismo', the disdain of gentlemen for industry and commerce and the desire rather to seek fame and fortune in war. There was, it is true, an urban Castile, the most illustrious example being Toledo itself, one of the great cities of early modern Europe. Nevertheless, the towns were enveloped by rural Castile which possessed both the physical and spiritual strength to impose its harsh and uncompromising values. In Castile that open and humanistic environment which nourished the great cities of Italy and the Rhinelands was lacking. The freedoms enjoyed there could not be tolerated in the religious and political climate of the Meseta. This condition was what Cajal described as the 'spiritual encystment' which converted the Pyrenees from a mountain range into 'a moral barrier.... a barrier of disdain'.[20] Throughout the period of Spanish hegemony, Castile remained its political centre. Even when the capital was moved from Toledo first to Madrid and then to the Monasterio del Escorial, Philip II's austere centre of government in the Sierra de Guadarrama, power remained firmly entrenched at the heart of the Meseta.

The centre of government was thus remote from the great centres of population and economic activity around the peripheries. Pre-eminent among these were Barcelona and Lisbon, the capital cities of Aragon and Portugal, and the great ports of the lower Guadalquivir. While continental Castile was able to achieve domination over these maritime peripheries, it was never in a position to make this domination absolute. Throughout the imperial period it was always 'Las Espanas' rather than 'Espana', and the non-Castilian provinces of the realm retained the Cortes and the Fueros, their traditional rights. Portugal, with a long tradition of independent nationhood behind it, remained a part of the Spanish Empire for only sixty years. Thus, while political, religious and military primacy rested in the centre, the balance of demographic, commercial and industrial power moved increasingly in favour of the periphery. At the beginning of the sixteenth century the population of Old and New Castile was a half of the Iberian total of eight millions. By the end of the century the population of Castile had actually declined while that of Iberia as a whole had risen by nearly a quarter.[21] This unfavourable demographic trend reflects a number of internal developments, including the expulsions of the Moors and the Jews who made up a highly industrious section of the population, high recruitment into the army, and an overall decline in intensive agriculture and its replacement by extensive pastoralism. In Castile, so it was said, 'the sheep eat the people', so leaving the land deserted.[22] The result of this was that Castile came to be filled with 'students, monks, beggars and bureaucrats' who were disinclined by nature and calling from productive endeavour. There was said to be 'a lack of people of the middle sort' who could have sustained a strong economy.[23] As a result, the centre of gravity of the country began to slide inexorably southwards towards trade with the New World which Spain was opening up and towards Andalusia, her own internal 'El Dorado'.

In the light of these centrifugal tendencies, the strict Catholicism of Castile, 'the redoubt of the true faith', became a weapon for the maintenance of political control. 'In proportion to its varieties and differences', said Trend, Spain needed 'a strong unifying force to weld it together and that force came from religion'.[24] It provided the justification for the dominant role of the Castilians, the 'Cristianos viejos', throughout both Iberia and Europe as a whole. 'The other parts of the Empire', observed Braudel, 'slipped imperceptibly into the role of satellites and Castile into that of the metropolitan power'.[25] Yet it was the measure of independence retained by these 'satellites' and their reluctance to place unlimited resources at the disposal of their hyperactive Castilian masters which forced the government to look beyond Iberia. This brought about what Elliott called 'the psychological crisis... which impelled it (Castile) into its final bid for world supremacy'.[26]

The first source of external resources had been the Aragonese trading empire and it was this which had led Spain into Italy. The 'threatened citadel' remained the most splendid prize which any European conqueror could hope to gain, bringing not only access to considerable wealth but also the prestige of the possession of Rome, the historic imperial and religious centre of the western world. The other potential source of wealth available to the Spanish monarch in Europe was the Netherlands. Since the late Middle Ages the composite delta of the Rhine had been the richest and most economically advanced region in northern Europe, and its cities had come to replicate the wealth and splendour of the Italian cities. The Netherlands, said Braudel, were becoming a European 'El Dorado' containing 'the treasures of the King of Spain, his mines and his Indies'.[27] By the second half of the sixteenth century they presented a marked contrast to Castile in religion, culture, politics and economic activity. The harsh and haughty Spanish masters came increasingly to be regarded as alien and unnecessary, and the discordance between the cultures of the inhabitants of this rich northern delta and those of the arid southern plateau was the geopolitical background to the Dutch War of Independence.

The final major source of Spanish wealth was, of course, the New World, and here Spain was able to maintain a dominant position through most of the sixteenth century. The economic raison d'être for the immense Spanish American empire was always the provision of the wealth needed for the pursuit of the country's hegemonial ambitions in Europe, a wealth which had the great advantage from the Spanish viewpoint of coming ready packaged as bullion. It had the further advantage of being of great value in proportion to weight and so could be shipped across the Atlantic in immense quantities. In the form of silver and gold coinage, it proved to be a means of exchange highly acceptable in the early modern economy of Europe, although its longer term effects were to be disastrously inflationary. While the New World was never responsible for more than a part of the total income of the Spanish Crown, the balance of income to expenditure was more favourable there than elsewhere in the royal dominions. The sheer availability of this source of wealth in the early sixteenth century made the tax burden on Spain's European possessions considerably less than it would otherwise have been. This was of particular importance for the Iberian periphery itself, which was therefore rather more willing to accept the demands of that ambitious Castilian foreign policy which sought nothing less than European and world supremacy.

Since the far-flung Spanish possessions, both in Europe and in the New World, were not territorially contiguous with Iberia, their maintenance was therefore dependent upon effective sea communications as well as overland routes. The only way in which the situation in Europe could have been ameliorated would have been through the defeat of France which lies strategically at the heart of western Europe and at the junction of its major land routes. The acquisition of territory from France would have linked Spain directly to the Lotharingian axis and so made her control over Italy and the Netherlands more secure. France, however, was sufficiently strong to resist being drawn into a Spanish dominated European imperial system, and in the following century was to become the major centre of opposition to it. The Spanish possessions formed a tight ring around France everywhere except to the north, and there was considerable Spanish influence even in England in the middle years of the sixteenth century. The line of movement from the Mediterranean through Italy to the Low Countries was therefore also from the French point of view a line of encirclement. The Spanish attempt to extend and consolidate this northwards into the British Isles was however not successful.

Spain's principal connecting link with her colonies in the New World became the lower Guadalquivir valley, and in particular the ports of Cadiz and Seville. That area combined possession of a fine natural routeway into the interior with some 200 kilometres of coastline facing the Atlantic. It was here that the principal build-up of Spanish naval power took place in order to protect the country's dominant maritime position. In the late sixteenth century Spain then redirected this maritime power to support her intervention in northern Europe. However, she was fundamentally a land power and her strategy remained a continental one. The principal function of naval power in the European sphere was seen as being the transportation of bullion and manpower - doubloons and tercios - from one territory to another. In the Mediterranean it had been used with considerable success and, with the support of the Genoese and the Venetians, maritime city states well used to naval warfare, successfully halted the westward expansion of the Ottoman Empire. This same naval power proved woefully inadequate however to cope with the unfamiliar physical environment of northern Europe, this inadequacy being made all too apparent in 1588 with the spectacular disaster of the Armada. The immediate cause of this was Spanish failure to deal adequately with the dangers of the northern seas, but in the longer term defeat was the result of the refusal of the north Europeans to bow down to Spanish hegemonial pretensions. It was in these northern waters, and subsequently in the Atlantic itself, that the Spaniards were to prove to be most vulnerable to indigenous naval power better adapted to its conditions.

By the end of the sixteenth century the attempt to sustain the bid for supremacy had become increasingly dependent upon a persistent and calamitous belief in gold and silver. Castile was becoming increasingly impoverished, the Iberian peripheries were restive and the northern 'El Dorado' had been transformed into Philip's 'damnosa hereditas', the effort to retain which was a constant drain on scarce resources. By the time of the death of Philip II in 1598, the Crown was massively in debt and his policies had come under considerable strain. Despite the pretensions to a global role, the basic interests of Spain remained in the Mediterranean, still the centre of the European world. Here, as principal standardbearer of the Cross against the Crescent, she had successfully halted the Turks and bound together the lands of the western Mediterranean as a functioning unit.

The fact was that Castilian ambitions eventually over-reached the limited capacity of Spain to translate them into reality, but her failure resulted also in large part from the profound changes which were taking place in the western ecumene as a consequence of the great discoveries. It is ironic, since Spain itself played so large a part in them, that they should have been largely responsible for the massive shift of wealth and power from the Mediterranean to north-west Europe. To Braudel, the great conflict of the sixteenth century was 'fundamentally a struggle for control of the Atlantic Ocean, the new centre of gravity of the world'.[28] It was above all the inadequacy of her indigenous physical and human resources which drew Spain westwards into 'that immense battlefield', the Atlantic Ocean, and there she expended her final energies as Europe's dominant power. Despite this period of dominance, Spain remained basically what she had been since the union of the Crowns, a power with an economically weak core region which had become progressively more debilitated as a result of the excessive demands placed upon it in the name of religion and national grandeur. Spain, said the Duc de Sully, is 'one of those states whose legs and arms are strong and powerful but the heart infinitely weak and feeble'.[29] In this respect at least the contrast with Sully's France could not have been more marked. Braudel viewed the problem in essentially geopolitical terms. Philip II's fundamental mistake, he said, had been 'not to go as far as possible to meet the flow of silver, to the Atlantic coast, to Seville or even better to Lisbon'.[30] In other words, the political dominance of the Castilian core region should have given place to that of the new economic core on the south-western periphery of Iberia.

The central contribution of the Castilian core was the maintenance of the ideals of the Reconquista which then gave Spain her raison d'être as a great power. Her failure was her inability to use effectively the vast resources from Europe and the Americas which for a time were at her disposal. The baleful effects of the reliance on gold and silver became evident as the economy declined at the same time as the supplies of bullion from across the Atlantic began to diminish. The truth of the old saying became ever more evident: 'Ponderoso caballero es Don Dinero'.

Braudel implied the existence of a functioning whole when he talked of 'the great web of the Spanish Empire' with the spider Philip II, 'El Prudente', at its centre in the gloomy Monasterio del Escorial.[31] Yet the three crucial strands of which it was woven – 'God, glory and gold' – failed to enmesh sufficiently to give it real durability. The centre of European power then moved definitively to the north of the Alps and the Pyrenees.

NOTES AND REFERENCES

1. V. Cornish, *The Great Capitals* (Methuen, London, 1923), p.97ff.

2. C. McEvedy and R. Jones, *Atlas of World Population History* (Penguin, Harmondsworth, 1978), pp. 134-7.

3. F. Braudel, *The Mediterranean and the Mediterranean World in the Age of Philip II* (Collins Fontana, London, 1972), p.661.

4. J.C. Dewdney, *Turkey* (Chatto and Windus, London, 1971), p.59.

5. V. Cornish, *The Great Capitals*, p.97ff.

6. F. Braudel, *The Mediterranean and the Mediterranean World*, p. 776.

7. H. Root, review article in *Comparative Studies in Society and History*, no. 20, 1978, p.628.

8. M.I. Newbigin, *The Mediterranean Lands. An Introductory Study in Human and Historical Geography* (Christophers, London, 1924), p.197.

9. J.C. Dewdney, *Turkey*, p.60.

10. H. Inalcik, 'The Rise of the Ottoman Empire', in M. A. Cook (ed.), *A History of the Ottoman Empire to 1730* (Cambridge University Press, 1980), p.41.

11. M.I. Newbigin, *The Mediterranean Lands*, p. 197.

12. F. Braudel, *The Mediterranean and the Mediterranean World*, p. 776.

13. N. Barber, *Lords of the Golden Horn From Sulaiman the Magnificent to Kemal Ataturk* (Macmillan, London, 1973), p.140. In 1529 Vienna was held by a garrison of only 15,000, while the army of Sulaiman the Magnificent was a quarter of a million. 'Sulaiman should have taken Vienna... had not

torrential rains prevented him from bringing up his big guns'.

14. F. Braudel, *The Mediterranean and the Mediterranean World*, p. 693.

15. F. Braudel, ibid., p. 781.

16. F. Braudel, ibid., p. 672.

17. This follows an established line of communication between the Mediterranean and northern Europe along which have grown up large commercial and industrial centres. Since the seventeenth century it has been central to the economic development of western Europe. The term 'Lotharingian axis' was coined by Nigel Despicht to indicate both its historical and its contemporary relevance. See N. Despicht, *The Common Transport Policy of the European Communities* (P.E.P., London, 1969).

18. The original core of Castile, as of Portugal, was in the Cantabrian mountains. Castile then united with the Kingdom of Leon and the capital was moved south to Burgos on the frontier with Arab Spain.

19. K.A. Wittfogel, *Oriental Despotism. A Comparative Study of Total Power* (Yale University Press, New Haven, 1957), p.426.

20. Quoted in: J.B. Trend, *The Civilisation of Spain* (Oxford University Press, London, 1960), p.143.

21. F. Braudel, *The Mediterranean and the Mediterranean World*, p.677.

22. Wittfogel disagreed with this seemingly plausible assertion. To him the cause of rural depopulation was 'the replacement of labour-intensive irrigation farming by labour-extensive cattle breeding'. See K.A. Wittfogel, *Oriental Despotism*, p.218.

23. J.H. Elliott, 'The Decline of Spain', in C.M. Cipolla (ed.), *The Economic Decline of Empires* (Methuen, London, 1970), p.185.

24. J.B. Trend, *The Civilisation of Spain*, p.82.

25. F. Braudel, *The Mediterranean and the Mediterranean World*, p.677.

26. J.H. Elliott in C.M. Cipolla, *The Economic Decline of Empires*, p.197.

27. F. Braudel, *The Mediterranean and the Mediterranean World*, p. 674.

28. F. Braudel, ibid., p.678.

29. C. Petrie, *Don John of Austria* (Eyre and Spottiswoode, London, 1967), p.280.

30. F. Braudel, *The Mediterranean and the Mediterranean World*, p. 678.

31. F. Braudel, ibid., p. 1236.

THREE

EMPIRE AND DOMINATION NORTH OF THE ALPS

As Spain went into decline as an imperial power during the early seventeenth century, Austria became in Braudel's words the 'foremost rampart of Christendom'.[1] The origins of this state, which from then on until the early twentieth century was to play so crucial a role in the affairs of Europe, lay in the eastern marchlands of the Holy Roman Empire. These constituted a wide zone from the Baltic to the Adriatic and were both the eastern boundary of the lands of the German people and the eastern frontier of Christendom facing the heathens. It was on this dynamic frontier of a militant and proselytising German Christianity that the Ostmark was established in the tenth century as a military outpost of the Duchy of Bavaria, one of the five stem duchies which made up the German realm. The Ostmark expanded its territory eastwards and became sufficiently strong to be made into a dukedom in its own right and in time to proclaim its independence of Bavaria. The historic core of the duchy was in the middle Danube valley, and its capitals were successively Pochlarn, Melk and Tullin, each further down the Danube than its predecessor. In the twelfth century the capital was once more changed, this time to the stronghold of Vienna, well inside what had been Slav territory. Vienna thus became what Cornish described as 'the advanced headquarters of a German frontier state'[2] poised for further eastwards advance down the Danube.

Since the structural alignment of central Europe is from west to east, the Imperial marchlands cut across it almost at right angles and consequently there were few natural obstacles to penetration eastward. The political structures which were established along this physically indeterminate and economically poor frontier zone were originally remarkably similar to one another. They had the status of 'marks' – frontier bastions – within the Holy Roman Empire, and were dominated by military aristocracies which based their power on the great estates carved out of land originally seized from the Slavs. They worked closely with the Church and with the militant religious orders whose purpose was to advance the frontiers of Christendom triumphantly eastwards.

While its political structure and raison d'être was originally much the same as the other marks, Austria was better endowed physically than most of those to the north, and by the thirteenth century had grown into a large and powerful state dominating the south-eastern corner of the Empire. A decisive event in its fortunes came in 1273 when Rudolf of Habsburg, a Swiss aristocrat, was elected Holy Roman Emperor and shortly afterwards acquired the duchies of Austria and Styria. Carinthia and Carniola to the south were soon added and, from then on, these mountainous lands around the fringes of the Empire became the territorial base of that Habsburg dynasty from which almost all the Emperors were subsequently drawn. The eastern frontier of the Holy Roman Empire soon stabilised along a line just east of Vienna, oriented from north-east to south-west from the Danube to the northern end of the Adriatic. Facing Austria on the other side of this frontier lay the large and powerful Kingdom of Hungary. Nomadic and warlike in their origins, the Magyars had for many centuries been building an imperial state of their own extending outwards from their centre in the Pannonian basin. The existence of this state set limits to further Austrian advance eastwards.

It became the policy of the imperial princes that, as Fisher put it, 'the Emperor should be pushed into the utmost corners of the Reich and closely occupied with non-German ambitions'.[3] As a consequence both of his being increasingly unwelcome as a participant in internal affairs and of the real dangers which lay beyond the frontiers, the activity of the Emperor came increasingly to be concentrated around the south-eastern peripheries. Nevertheless the principal centres of commercial power in the Empire remained in the west and north, especially in the string of prosperous cities following the Lotharingian axis and in the ports of the Hanseatic League along the Baltic and North Sea coasts. The early Habsburg Emperors, nominal rulers of an increasingly centrifugal state, did initially make an attempt to extend and consolidate their power inside their enormous domain. However, in the fifteenth century the European world was profoundly shaken by the Ottoman invasions, and this soon became a matter of overwhelming concern to the Emperor, the 'primus inter pares' of Christendom and holder of its 'foremost rampart'. Following their defeat of the Serbs, the Ottomans then streamed across the Balkans and on into the Pannonian basin. In 1526 at the Battle of Mohacs they defeated the Hungarians and occupied most of their kingdom. The Crown of what remained of Hungary was offered to and accepted by the Emperor Charles V, and Austria found herself once more on the front line of an embattled Christendom. In 1529 Vienna was itself besieged, but the Turks were forced to retreat and regroup in the Pannonian plain. They then took up positions some 50 to 100 kilometres to the east of the Imperial frontiers and this became the 'Militärgrenze' across which for the next one hundred and fifty years the principal guardian of Christendom faced the self-proclaimed champion of Islam. The failure of the final desperate assault on Vienna by the Turks in 1683 was followed by a general Austrian advance eastwards. Within ten years Hungary had been liberated and the Habsburgs had claimed their full royal inheritance. In the Treaty of Carlowitz in 1699 the Ottomans accepted the loss of Hungary and Transylvania, and further campaigns subsequently pushed them back south of the Danube. In his triumphant advance to the east, the Emperor was supported and encouraged by many of the other princes of Christian Europe, but only in a desultory fashion by those of the Holy Roman Empire itself. This marked the beginning of the detachment of the Habsburg imperial role in their own dominions from their older religious role as Holy Roman Emperors. The process was not finally completed until 1804 when the Holy Roman Emperor Francis II was proclaimed Emperor of Austria, shortly before the Holy Roman Empire was itself dissolved. In 1713 the Austrian Habsburgs added further to their secular power with the acquisition of the Spanish Netherlands, thus gaining an important foothold in north-west Europe. A little over half a century later, as one of the three participants in the Partitions of Poland, Austria also acquired substantial territory to the north of the Carpathian mountains. She then moved into the role of hegemonial power over that part of Italy north of the Po and east of the Piedmontese frontier. As a result of this expansionist activity, by the late eighteenth century Austria was the dominant power in the centre of Europe from the middle Danube westwards to the Rhine and from the North European Plain to the Adriatic. In geopolitical terms she had become the hegemonial power of an extended Mitteleuropa[4], and more than any other state came to typify that 'ancien régime' which, until the French Revolution, was to dominate most of continental Europe from the English Channel to the Urals. Thus, barely a century after the final siege of Vienna and the definitive removal of the Ottoman threat, Austria had been transformed from being the personal patrimony of the reigning house of the Holy Roman Empire into a large and impressive imperium in its own right. This extraordinary process will now be examined from a geopolitical perspective.

The source of the Danube lies in the Schwartzwald, only forty kilometres from the Rhine, while its delta on the Black Sea, 2,000 kilometres to the east, is shared between Russia and Romania. It is therefore the most important natural link between west and east in Europe. Its long course divides into three basins each separated from the others by a belt of higher land through which the river has incised itself. Each of these three basins has become the homeland of a particular people. In the upper basin there are the Germans, in the centre the Hungarians and, closest to the Black Sea, the Romanians. Around the southern fringes of the middle and lower basins there is a considerable South Slav presence, although the heartland of the South Slav peoples is in the Balkan mountains to the south. This mighty river was the central axis of the Austrian Empire and the middle basin, the Pannonian plain, was its principal focus of internal communications.

The historic core of the Austrian state is in the hills and mountains which lie between the upper and middle basins. The eastern boundary of this core region, and of the Holy Roman Empire itself, followed the fringes of the mountains around the western edges of the central basin. The historic frontier of the Duchy of Austria followed the river Leitha, a south-bank tributary of the Danube to the east of Vienna. After the defeat of Hungary by the Ottomans the frontier of the Habsburg domains was moved forward into the Pannonian plain. The historic geopolitical boundary between the spheres of the powers of central and eastern Europe was then per- petuated here as the 'Militargrenze'. In the process of rolling back the Turks in the late seventeenth century the Austrians then crossed this frontier and for the first time brought the whole of the central basin under their command. Austria had thus become the principal Danubian power and the successor state to the Ottoman Empire in south-eastern Europe. In Toynbee's words, 'from the moment of the Danubian Habsburg Monarchy's foundations, its fortunes followed those of the hostile power, whose power had called it into existence, in each successive phase'.[5] The fate of the Hungarians was to be transferred from one imperial allegiance to another, but while they had been on the northern frontiers of the Ottoman Empire they were now, fatefully, located at the geographical centre of Austria's new Danubian empire. However, despite immense pressure from their new imperial masters they re- tained sufficient strength and cohesion to resist the kind of absorption which was to be the fate of many of the Slav peoples when faced with overwhelming German power. This resistance was to have profound long-term consequences for the nature and durability of the Austrian Empire. One of these consequences was that the political centre of the Empire remained entrenched in the historic core rather than being moved eastwards into the new geographical centre in the Pannonian basin. Another was that the old geopolitical front- ier now became a significant internal division within the Empire between the German lands in the west which were within the Holy Roman Empire and the Magyar-Slav lands in the east which never became a part of it. 'Cisleithania' was to the Austrians their real homeland, essentially European in culture and history, while 'Transleithania', even after so long

Figure 3.1: The Geopolitical Centre of the Austrian Empire

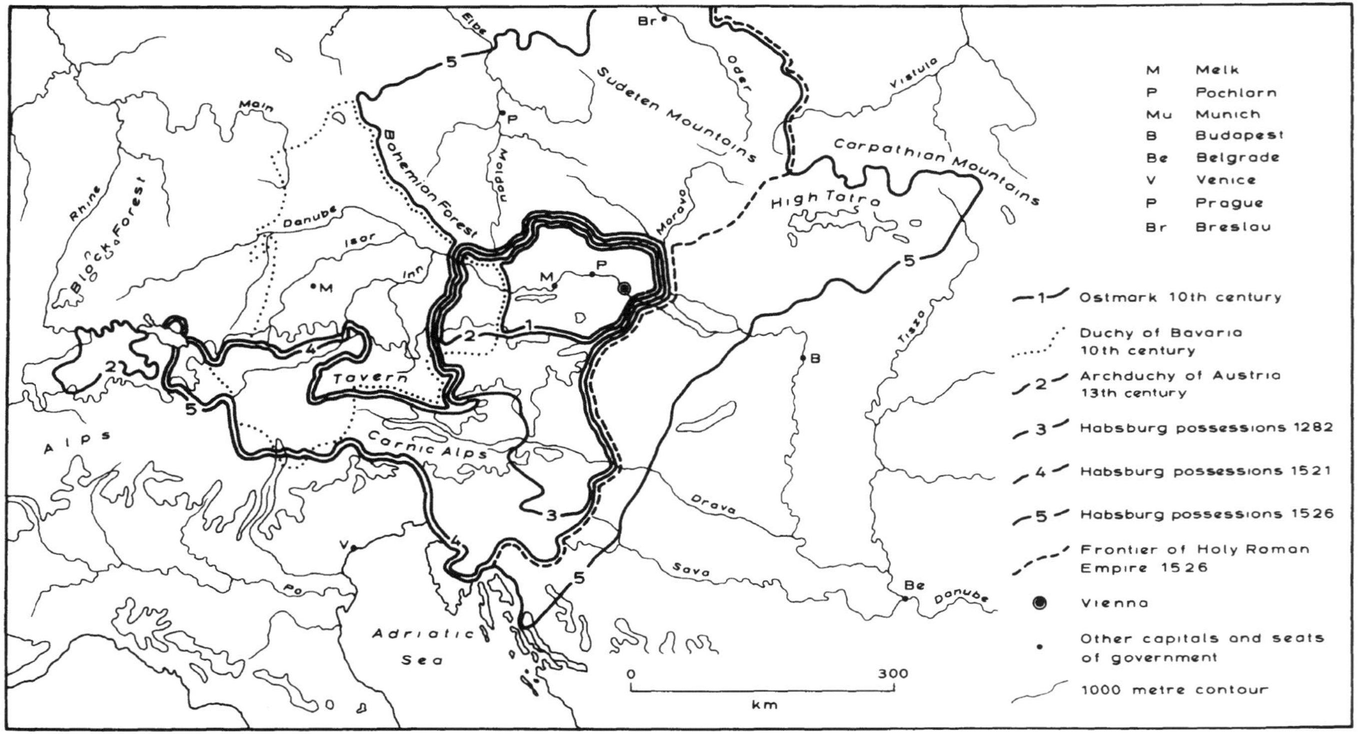
M Melk
P Pochlarn
Mu Munich
B Budapest
Be Belgrade
V Venice
P Prague
Br Breslau
1 Ostmark 10th century
Duchy of Bavaria 10th century
2 Archduchy of Austria 13th century
3 Habsburg possessions 1282
4 Habsburg possessions 1521
5 Habsburg possessions 1526
Frontier of Holy Roman Empire 1526
Vienna
Other capitals and seats of government
1000 metre contour
0
300
km
Elbe
Main
Rhine
Black Forest
Danube
Isar
Inn
Bohemian Forest
Moldau
Sudeten Mountains
Oder
Morava
Vistula
Carpathian Mountains
High Tatra
Tisza
Drava
Sava
Po
Tavern
Carnic Alps
Alps
Adriatic Sea

an association, was seen as being an outlier of Asia and a colonial territory.[6] Since Austrian influence also extended westwards to the Lotharingian axis, she retained an interest in the Mediterranean and western Europe. However, her hold in these areas was variable and precarious and of a different order from that in her eastern possessions. Vienna was thus in reality the fulcrum of two overlapping and contrasting empires, the one historical, religious and centrifugal and the other modern, secular and bound together in allegiance to the Habsburg dynasty. In the former they reigned by grace of the Imperial princes; in the latter they ruled by right of conquest. Despite her hegemony throughout Mitteleuropa, it was the east which became the focus of Austria's political ambitions. However, neither it nor any other part of her Empire was to prove adequate to sustain the country's hegemonial position. She conquered no El Dorado which could be turned to this purpose. Neither was she successful in controlling a rich periphery of commercial cities. Italy, which had fulfilled this role for Spain in the sixteenth century, was by the eighteenth century in a state of considerable decline. The economic centre of Europe had moved north-westwards, but even the once rich Austrian Netherlands, which Braudel had seen as having not long before been part of 'the treasures of the King of Spain, his mines and his Indies', was now totally overshadowed by the pre-eminence of the Dutch and the British.

While the eastern thrust of her imperial drive took Austria away from the rich maritime world, at least it took her into one in which she was the dominating power. However, even here as has been seen, there were strict limits to the extent of her control, and these arose from the continuing importance of the internal geopolitical frontier. When Habsburg Austria first emerged as an imperial power, her total population was some seven millions, well over half of it being German. By the late eighteenth century the total population of the Habsburg Empire had risen to some 25 millions, but by then only a quarter of it was German.[7] The largest single non-German section of the population was Slav, and the Slav peoples felt their own strong sense of nationality, based on language, history and culture. It proved impossible for nationalism of this sort to coexist with the dynastic, multi-national and German-dominated structures of the Austrian Empire. The situation was made more difficult by the fact that in the Pannonian basin at the geographical heart of the Empire lived the Magyars who possessed both their own strong sense of national identity and also the economic and demographic strength to maintain it. As has been seen, historically, Magyar power had impeded Austrian penetration into and control over the middle and lower Danube, and now it was they who lay astride the Empire between the Austrians and their Slav subjects, and who came to constitute an important alternative focus of power and influence within the Empire. It is ironic that while the title to the Crown of Hungary had added legitimacy to the Habsburg thrust to the east, it was this same crown which most strongly resisted the completion of Austrian domination. This situation derived not only from the build-up of Magyar nationalism in the nineteenth century, but also from Hungary's own historic position as the imperial power of the middle Danube region. The Hungarian uprising of 1848 gave warning of a willingness to disrupt the multi-national Empire of the Habsburgs and, as a direct result, the Ausgleich of 1867 elevated the Hungarians to the status of imperial co-partners. This at last signified equality between 'Cisleithania' and 'Transleithania', although it had the effect of strengthening the internal geopolitical frontier.

The Ausgleich was the beginning of the end for Austrian dominance. In the previous year her defeat by Prussia at Königgratz-Sadowa had finally removed her from her hegemonial position in Germany. Thus within two years she was removed from her hegemony both in the western and the eastern parts of her extended sphere. The same consciousness of growing enfeeblement which had forced the Austrians to make concessions to the Hungarians then drew them into an alliance with the powerful new German Empire. While this renewed the strength and cohesion of Mitteleuropa as a whole, Austria-Hungary was now relegated to a subordinate position within it. The problems arising from the multinational character of the Empire were by no means solved and they were aggravated by one final act of military expansionism. This was the annexation of Bosnia in 1908 which served to increase the hatred of the South Slavs for their imperial masters. This came to a head eight years later with the assassination of the Archduke Franz Ferdinand, heir to the Imperial throne, in the Bosnian capital of Sarajevo. The war which ensued led to the dissolution of the great empire which had presided for so long over the destinies of central Europe.

During the second half of the seventeenth century, France moved into the role of leading power in the western part of Europe. Her rise to a dominant position was neither as rapid nor as spectacular as had been those of Austria or Spain. She had been closely involved in the affairs of western Europe since the days of Charlemagne, an involvement which arose naturally from the country's highly central location. Its territory forms the western terminus of trunk Europe where the continent's principal macroregions approach the sea. Here these macroregions, so precisely defined in Mitteleuropa, gradually break up into gentler and more varied landscapes. Nevertheless, the north European lowlands, the Hercynian mountains, the Alps and the variable relief features of the Mediterranean coastlands can be clearly identified in French

Figure 3.2: The Geopolitical Centre of France

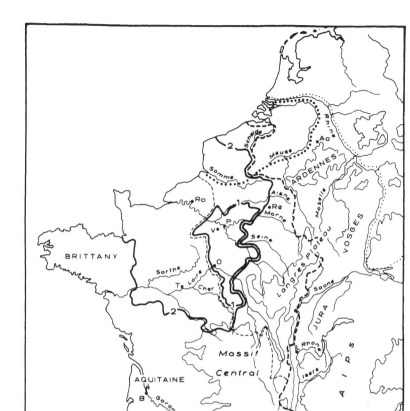

territory. The east to west alignment of central Europe is here replaced by a north-east to south-west one as the physical belts swing around to reach the sea between the Bay of Biscay and the Alps. While to the north, west and south the boundaries of French territory are clearly defined, to the east they merge into the physical features of central Europe. The macroregions are linked together by the major river valleys, most of which flow towards the west and north-west and so form natural routeways to the Atlantic and the North Sea. The one significant exception to this is the Rhône which flows southwards into the Mediterranean.

The French section of the North European plain consists of two large hydrographic basins, the Seine and the Garonne. Between them flows the Loire, the smaller basin of which provides a natural link between them. The Ile de France, the historic core region of the French state, is located in the lowlands between the Seine and the Loire (Figure 3.2). In the east the basin of the Seine adjoins that of the Rhine, the largest hydrographic basin in western Europe. This river had been the eastern frontier of the Roman Empire and the provinces of Gallia (Gaul) extended westwards from the Rhine over most of the territory of present day France. Charlemagne for a time unified the territory of Roman Gaul with the Germanic lands of the Rhine basin, but this unity was to be short-lived and his empire disintegrated within half a century of his death. The imperial title was for a time held by the monarch of the Middle Kingdom, Lotharingia[8], which consisted of a relatively narrow zone extending from the North Sea to the Alps but included the whole of northern Italy. The principal centre of power, and in the following century the imperial title itself, moved eastwards to Germany. This Holy Roman Empire of the German People rapidly became a large and expansionist state which arrogated to itself primacy in Christendom as heir to Rome. It was this state which made that drive to the east of which the Ostmark was a part.

During this period of German primacy, the eastern frontiers of France were tightly drawn along a line stretching from the Schelde to the Meuse, across the Langres plateau and then west of the Rhine basin to the Mediterranean. Virtually the whole of the basins of the Rhine and the Rhône at this time lay within Imperial territory, and medieval France was only two-thirds the size of Roman Gaul. Indeed, by the twelfth century the Empire was three times the size of France, which itself was no more than a loose feudal conglomerate, with the King as 'primus inter pares' and the real power in the hands of his principal vassals. For much of this period France constituted a power vacuum, its rich agricultural land and commercial cities making it a delectable prize for such peripheral predators as the Normans, the Angevins and the Burgundians. Effective royal power dates from the thirteenth century when Philip Augustus and his immediate successors began to extend the royal domain out from the Ile de France over almost the whole of the Paris and Aquitaine basins. The Angevin claims to the French throne, based on the feudal status of their Norman predecessors by virtue of the lands they held as vassals of the French Crown, culminated in the Hundred Years War and, by the middle of the fifteenth century, the expulsion of the English. The growth of resistance to this threat to French independence came to be personified in Joan of Arc, and contributed to the emergence of a precocious sense of national identity. After this experience, France was rapidly transformed into a powerful country whose monarchs, until then preoccupied with the consolidation of their own internal power, began to look outwards onto the European stage. The invasion of Italy by Charles VIII in 1492 was an early manifestation of this, but the fulfilment of such aspirations had to await the decline of Spain. As has been seen, Spain had secured a dominant position in Italy and along the Lotharingian axis, and while this was intended principally to link together her dominions in southern and northern Europe, it also formed a steel girdle which held in France as tightly as she had formerly been held in by the Empire.

The growth of French power in the seventeenth century was founded on a large and diverse national territory which possessed substantial agricultural and industrial resources. Its population had by then risen to 20 millions, larger than that of any neighbouring state, so enabling Louis XIV, 'Le Roi Soleil', to field huge armies in the pursuit of his territorial ambitions. The centralisation of the French state, founded on the dominant position of Paris, was reinforced by the policies pursued successively by Cardinal Richelieu and Louis XIV. To the position of Paris as centre of communication and commerce at the heart of the country's largest and most fertile river basin, was now added the increasing centralisation of political power on the capital. This centralisation gradually eroded the powers of those centrifugal elements in the state such as the nobility and the provincial parlements, and thus put the national resources more directly under royal command. This domination of the capital over the provinces and of the north over the south led to the phenomenon of 'Paris et le désert francais'.[9] The centripetal forces operating within the French state were also the basis of the linguistic and cultural dominance of Paris which overwhelmed the regions and produced a level of homogeneity throughout the national territory far greater than that to be found in either Spain or Austria. This physical and cultural strength of the French core region in the Paris basin was a marked contrast to that 'heart infinitely weak and feeble' which the Duc de Sully had found in Spain. The powerful culture which was generated by Paris and sustained by the physical and human resources of the provinces was the 'civil-isation francaise' which dominated Europe during the eigh-teenth century. It became the principal vehicle for the Enlightenment which had so profound an effect on the culture of the nations of north-west Europe. This was to provide both a strong basis and a powerful justification for the hegemonial ambitions of the French.

The territorial issue which was considered to require the most urgent attention during the second half of the seven-teenth century concerned the vulnerable eastern frontiers. With the exception of the acquisition of Dauphine and Provence in the lower Rhone area, these had altered very little in seven hundred years. Yet they had few geographical justifications and were generally regarded as being unsatis-factory and insecure, deriving from a time when France had been under sustained external pressure. In the seventeenth century the whole situation changed as France moved into its new role of being itself a powerful and expansionist state. The initial military thrust of Louis XIV was northwards into the Netherlands. The twin geopolitical attractions of this were the acquisition of a more favourable frontier and control over the rich commercial centres of the Rhine delta. By this time it was the Dutch who had become Europe's major commercial power and, for the second time within a century, they were forced to defend themselves against a continental power desirous of wresting their hard-won freedom from them. However, water was again as much a friend of the Dutch as it was the enemy of the French and the Spaniards before them. The combination of superior Dutch naval power with the opening of the dykes to impede their military advance, de-feated the territorial ambitions of the French in the Rhine delta. This deflected French attention to the eastern frontier, and here greater success was attained. The campaigns culmi-nated in the capture of the fortress city of Strasbourg in 1681 and from then on this city became the symbol of the French presence as a riparian power on the Rhine.

The considerable eastward movement of the French frontier which resulted from these conquests underlay a substantial shift in the balance of power from the east to the west of the Rhine. One of the many arguments then used to justify this considerable expansion was that of 'les limites naturelles'. In the intellectual climate of seventeenth century France this consisted of something far more rational than simply that evocation of the 'magical significance of mountains and rivers' which Kolarz saw as being attributed to natural frontiers by the twentieth century.[10] It was 'a simple and convenient notion', said Ancel, which arose from 'the need for order which dominated the French monarchy'.[11] The natural frontiers of France were considered at the time to be 'the Seas, the Alps, the Pyrenees and the Rhine'.[12] The contention that the Rhine itself came into the category of a natural frontier has been widely attributed to Cardinal Richelieu who was at the helm of the French state in the middle years of the seventeenth century and who laid the foundations of the country's later dominance. However, it has been shown that the idea was actually a much older one than this and can be traced back to Rome and to Caesar's statement that 'Populi Romani Imperium Rhenum finire'.[13] The river was not only the eastern frontier of Roman Gaul, but also the limit of the 'Romanitas', the area of Latin civilisation in Europe. When the idea was reactivated in the seventeenth century, contemporary arguments of nature and natural law were used to supplement the historic precedents. This entailed the belief that there was a 'natural' France, the outline of which, in the words of the French historian Sorel, 'avait été dessinée par la géographie'. [14] It was this concept which was now to be made into a reality on which the country's strength and security was to rest. A parallel concept was also formulated of an ideal geography founded on Cartesian principles. Physical reality and abstract concept were to be brought together at all levels by the subduing of nature and the creation of 'paysages humanisés'. The imposition of order on disorder, of geometrical purity on the untidy facts of physical and political geography was the central desire of an age when Louis XIV was the focus and inspiration of an increasingly secular state religion, and his palace at Versailles represented the triumph of logial creativity and of geometrical harmony.

Despite all this, by the time of his death in 1714 Louis XIV had been only partially successful in his military ambitions and the Rhine frontier had been only imperfectly secured. The second and far greater advance to the east came some eighty years later and was a product of the immense national energies released by the French Revolution. In 1792, in the name of 'liberté, egalité, fraternité', the French advanced into the territory of their eastern neighbours. At the same time they were well aware of the historic precedents which they were quite happy to use in support of their ideological arguments. The expansion of the country to 'les limites anciennes et naturelles' was justified by Carnot in terms of natural law. This gained the active support of Danton, who asserted that those frontiers 'marked out by nature' were vital to the nation's strength and security.[15] As a result of the French Revolution, said Ancel, the notion of the frontier entered into the moral domain.[16] Vidal de la Blache's 'esprit frontalière' had already found political expression at Strasbourg in 1790 in the sign facing eastwards which declared 'Ici commence le pays de la liberté'. When the Revolution then spilled over the frontiers of the land which had given it birth, its ideas were borne eastwards into Mitteleuropa with the victorious armies of the Republic. The river Rhine then ceased to be thought of as a natural frontier and was transformed into a link between western and central Europe. It became easy for the Revolutionaries to believe, without giving the matter too much thought, that France herself had a 'destiny' in the Rhinelands.[17] The first task in the accomplishment of this destiny was the incorporation of the ancient lands of 'la civilisation rhénane', and the granting to them as of right those benefits which the Revolution was believed to have given to France herself.

The subsequent internationalisation of the ideals of the Revolution were used to justify the assault on the 'ancien régime' over a far wider area of Europe. This took Napoleon across Europe from victory to victory. In 1804 he was proclaimed Emperor, and two years later the Holy Roman Empire was brought to an end after a thousand years. Between the east bank of the Rhine and the basin of the Elbe, Germany was united into the Confederation of the Rhine, a dependent state of France.

By 1810 the French Empire itself extended well beyond any 'limites naturelles' previously conceived of. In northern Europe it comprised all the territories west of the river Rhine and extended beyond the Rhine delta as far as the Elbe estuary. The Dutch, formerly the centre of resistance to French expansion, had for long been in decline and were no longer a political or economic force to reckon with. Unable to resist the overwhelming might of Napoleon's armies as they had so successfully resisted those of his predecessors, they had been defeated and reduced to the subordinate status of the Kingdom of Holland within the French Empire. Piedmont, Florence, Rome and the Illyrian province across the Adriatic were also incorporated, giving the Empire a total area of 800,000 square kilometres and a population of 50 millions.[18] Beyond these bloated frontiers the rest of Italy, Germany, Spain and the Grand Duchy of Warsaw were its client states. In geopolitical terms, the French Empire had achieved a position of dominance over the whole of trunk Europe together with its two southernmost peninsular extensions.

The collapse of the Napoleonic Empire came swiftly and by 1814 it was all over. The French armies were defeated, the country occupied, the 'ancien regime' restored and Napoleon sent into exile. The Congress of the European powers meeting at Vienna declared the French frontiers to be those of 1789, thus confirming the gains made by Louis XIV but no later ones. France was to remain one of the riparian states of the Rhine but was definitively excluded from the Netherlands.

The most fundamental geopolitical factor underlying the failure of France to consolidate her position of dominance was her total inability to subdue the western and eastern peripheries of Europe and to gather them into her Francocentric system. The maritime periphery was under the control of Great Britain which by the late eighteenth century had become the world's most important commercial and naval power. Throughout the century France had attempted to dislodge Britain from her pre-eminent position here, and her failure to do so denied her access to the sort of El Dorado which had for a time been readily available to Spain. As Britain drew her ring of sea power ever tighter around Europe from the Atlantic to the Mediterranean, the continent became increasingly isolated from the rest of the world and thus dependent upon its own indigenous resources. The Battle of Trafalgar in 1805 set the seal on Britain's maritime supremacy, and after this France lacked any effective naval strategy.

Following Napoleon's victories at the battles of Austerlitz, Auerstadt and Jena, France had replaced Austria as the hegemonial power of Mitteleuropa. The French sphere then extended eastwards to Poland and Galicia and there, along an 800 kilometre frontier, she faced the Russian Empire. After the treaty of Tilsit between the two empires, the French increasingly came to regard Russia as being both an unreliable ally, reluctant to isolate Britain, and a potential rival of vast dimensions and considerable ambition. Consequently in 1812, after much preparation, the grande armée crossed the Nieman and thrust deep into the Russian heartland. Despite initial victories and the occupation of Moscow, the whole enterprise rapidly came to a disastrous end and only a fraction of the grande armée succeeded in returning to France. The French had unwisely ventured out of the confines of western and central Europe, built on a scale to which they were accustomed and with a physical environment with which they were familiar, and into an alien land of vast open spaces and a ferocious continental climate. It was here that the French, until then seemingly invincible in land battle, met with their first and most catastrophic defeat. As Napoleon shrewdly observed, the most formidable of all his opponents had been 'le général février'. The events of 1805 on sea and 1812 on land demonstrated conclusively that France was ill-equipped to cope with the extremes of either maritime or continental conditions. Great Britain and Russia represented alternative geographical centres of power within the western ecumene which France was in no position to subjugate. The brief dominance of France remained confined to the continent of Europe from the English Channel to the Baltic-Black Sea isthmus.

Ultimate military failure was also brought about by the inadequacy of France's indigenous and human resources to satisfy the massive demands made by its rulers. Despite the fact that by the end of the eighteenth century the population of France itself had reached 30 millions, a half as much again as under Louis XIV, the endless demands of Napoleon's 'levées en masse' proved to be seriously debilitating. It also had serious effects on the capacity of the economy to produce the essential foodstuffs and weapons of war which the country needed. Underlying this inadequacy was the bleak fact that France had already lost her economic pre-eminence. During the second half of the eighteenth century Britain was in course of that economic transformation brought about by the Industrial Revolution which was shortly to make her the world's most powerful industrial nation. Since the seventeenth century France had been seeking to achieve a commanding position in the Lotharingian axis, but ironically by the time they were successful in this, the lands from the Rhine to northern Italy had ceased to be the internal El Dorado which they once were. Toynbee saw them as having become 'a broken down cosmos' of city states in what was by then the 'decaying centre' of the continent.[19] With the constraints on maritime commerce imposed by the British navy, the material resources available to Napoleon thus became ever less sufficient.

In addition to the failure to subdue the peripheries and the relative weakness of the home base, a final factor in the French defeat was her failure to translate the ideals of the Revolution into an international system which the rest of Europe could accept, let alone welcome. The Revolution had given France a new internal political geography founded on eighteenth century rationalism. Historic provinces and eccentric medieval survivals were swept away and replaced by new 'départements', each of roughly the same size and neatly drawn on specific geographical principles. This system was also imposed on the newly incorporated territories up to and beyond 'les limites anciennes et naturelles'. However, what had been conceived of as a great new beginning and a total break with the 'ancien régime' began within a decade to take on in many ways the character of that which it had replaced. The new and rational political geography became the territorial basis for still greater centralisation on Paris and a means of ensuring that the resources of the Empire could be more effectively deployed in the service of the French state. This also became the principal motive for French actions in the satellite states to the east of the Rhine and south of the Alps where the crusade against the evils of the 'ancien régime' took on a more chauvinistic colour. The French, who came as liberators, stayed to become conquerors, and finally erected their own empire on the ruins of those they had so recently overthrown. The forces of reaction were thus enabled to return in the guise of liberators and the overthrow of French rule was greeted with widespread relief a mere quarter of a century after the Revolution.

France had emerged in the second half of the seventeenth century as the most powerful European state just at the time when the centre of power was moving definitively from Cisalpine to Transalpine Europe. She possessed the considerable geographical advantage in pre-industrial Europe of a productive and varied agricultural base which was able to sustain a far greater population than any neighbouring state. She also had a powerful and dominating core region which was able to marshal the resources of the state far more effectively than had been possible in centrifugal Iberia. While 'Las Españas' had remained plural, 'La France' was singular and was made ever more so by the centralising policies of successive rulers. The Revolution was primarily a product of the French capital and its core region. It was founded on the principles of natural law, logic and moral good, and it was held to be axiomatic that France was its natural homeland and Paris its intellectual and spiritual centre. As a result, the frontiers of France, as Ancel observed, became as much moral as political phenomena. However, this idealism soon gave place to materialism and one of the expressions of this was the creation of an empire, the vehicle for securing and perpetuating a position of dominance. Carrying the trappings of Rome and the blessing of the Pope, it was proclaimed just as the Holy Roman Empire was about to come to an end. Thus it became in many ways another resurrection of the Imperium Romanum, this time transferred from the banks of the Tiber to those of the Seine.

France, while still official standardbearer of the ideals of the Revolution, was thus revealed in a far more ambivalent role. She was now at one and the same time both the bearer of Enlightenment and the defender of established values. It was because of this contradiction, said Toynbee, that 'the Napoleonic Empire carried within itself the Promethean seeds of its own inevitable failure'.[20] But related to this there was also the failure to subjugate the two powerful peripheries which were then in a position to give aid to those forces throughout Europe which aimed to overturn French dominance. It was the combined strength of these two which eventually put an end to the French dream, and in so doing succeeded in reimposing the 'ancien régime' upon a reluctant Europe. With the defeat of France these two peripheries became so powerful and so global in their reach that for half a century the European continent lay in their shadow.

A little over half a century following the events which brought about the end of the French bid for supremacy, the army of the French Emperor Napoleon III was routed by the Prussians at Sedan just south of the Ardennes. In 1871 the second Deutsches Reich was proclaimed at Versailles, and the King of Prussia became its first Emperor. Until then France had continued to cherish the hope of regaining at least some of her former glory, but the previously dominant state was soon to be eclipsed by the arrival of this newest addition to the ranks of Europe's great powers.

As with the other German power, Austria, the origins of the Prussian state lay in those marchlands of the Holy Roman Empire where the Germans, in the name of Christendom, did battle with the heathens to the east. The Altmark, its original core, was founded in the tenth century on the west bank of the Elbe by Henry the Fowler, Duke of Saxony. He was the first of the powerful Saxon line of Emperors which embarked upon a revival of the Empire and brought about an impressive territorial expansion.[21] The Altmark was intended to be instrumental in the expansion to the east and during the following centuries its territory grew eastwards to the Oder and beyond. The conquered lands became Mittelmark as far as the Oder and Neumark between the Oder and the Warthe. The whole then took the name of Brandenburg, from the old Wendish stronghold of Brannibor west of the Havel lakes in Mittelmark, and this remained its capital until the fifteenth century when it was transferred eastwards to Berlin, at the junction of the Havel and the Spree. Its major physical axis followed a glacial depression, or urstromtal, from the Elbe to the Oder, and most of the mark consisted of glacial lowland country - 'sour sandy soil on which brutal manorial chiefs and brutish serfs fought a dour and relentless battle with nature and with each other'.[22] Located in the middle of the North European Plain, the mark had neither a sea coast nor any clearly defined limits to its territory. The principal concern of the military landowning aristocracy which ruled this bleak marchland was the conquest of more land, and it was assisted in this objective by a Church which aimed to convert the heathens and to establish new bishoprics. In pursuit of these aims Germans from the west were encouraged to move eastwards as colonists.

The other historic core of the Prussian state lay in the coastal lands between the Vistula and the Nieman which were wrested from the native Slavs by the Teutonic Knights. This fierce Christian order had originally been engaged in crusading in the Holy Land, but it was subsequently commissioned by the Church to join with other Christian orders to convert the eastern heathens. In fact, 'conversion' took the form of military conquest of the lands east of the Baltic Sea from the Vistula to the Gulf of Finland, a process which was not to halt until the fifteenth century when the Knights were finally defeated at the Battle of Tannenberg by the Polish-Lithuanian confederacy, and for a time became vassals of the Polish monarchy. By this time Brandenburg itself had become an electorate and was ruled by the Hohenzollern dynasty which in the early seventeenth century also succeeded to the throne of Prussia. From that time on the fortunes of these two territories were inextricably intertwined, and by the eighteenth century Brandenburg-Prussia, renamed the kingdom of Prussia, had become one of the acknowledged great powers of Europe. The persistent problem of the physical discontinuity of the two territories which made up the state was finally solved by Prussian participation in the dismemberment of the once large and powerful Polish state which had called a halt to the German eastwards expansion. In the three Partitions of Poland which took place in 1772, 1793 and 1795, Prussia acquired the whole of the middle and lower Vistula basin from Warsaw to the Baltic, and also the basin of the Warthe, an eastern tributary of the Oder. This, together with the earlier acquisition of Silesia from Austria, greatly increased the size and cohesion of Prussian territory, and by the end of the eighteenth century she had a total area of around 200,000 square kilometres and a population of 9 millions, almost as great as that of England at the time. Apart from the Altmark itself, the whole of the state lay east of the old Elbe frontier, and represented the result of conquest over many centuries.

Despite it enormous increase in size, the state still consisted mainly of glacial lowlands traversed by the lower courses of large northwards-flowing rivers. It was this bleak and unpromising geography which prompted Voltaire's scornful comment to the Prussian King Frederick the Great that, despite all his pretensions to grandeur, he was still only 'le roi des lisières'.[23] However, by that time this had ceased to be altogether true, since the seizure of Silesia from Austria enabled Prussia for the first time to break out of the North European Plain and to add the middle Oder basin and the Sudeten mountains to her territory. The acquisition of Silesia, thought of by the rest of Europe as being a wilful and even random act of brigandage, in fact made a vast difference to the strength of the Prussian state. Its size was thereby increased by a half, its population doubled, its agricultural potential greatly increased by the rich löss soils of the Börde zone, and valuable mineral resources were acquired in the Mittelgebirge to the south. It was the possession of this Silesian treasure house which finally gave Prussia the potential to be a great power. The other acquisition of immense significance was a coastline on the southern Baltic with such important ports as Stettin and, following the Second Partition of Poland, Danzig. Although the population of this enlarged Prussia remained predominantly German, the Partitions added a Polish element in the east which began to give Prussia an imperial aura which, unlike Austria, it had not previously evinced.

By the late eighteenth century Prussia had become physiclly defined by the flanks of the North European Plain in the south and the Baltic coast in the north (Figure 3.3). A major strategic problem was that it still lacked defensible frontiers to the west and east and thus remained particularly vulnerable from these directions. The extent of this vulnerability was demonstrated during the Seven Years War when the existence of the state was threatened by a hostile coalition of Austria, France and Russia. Half a century later the massive defeat by Napoleon's 'grande armée' at Jena for a time completely eliminated Prussia as a power and its resurrection was only brought about by the victory of the

Figure 3.3: The Geopolitical Centre of Germany

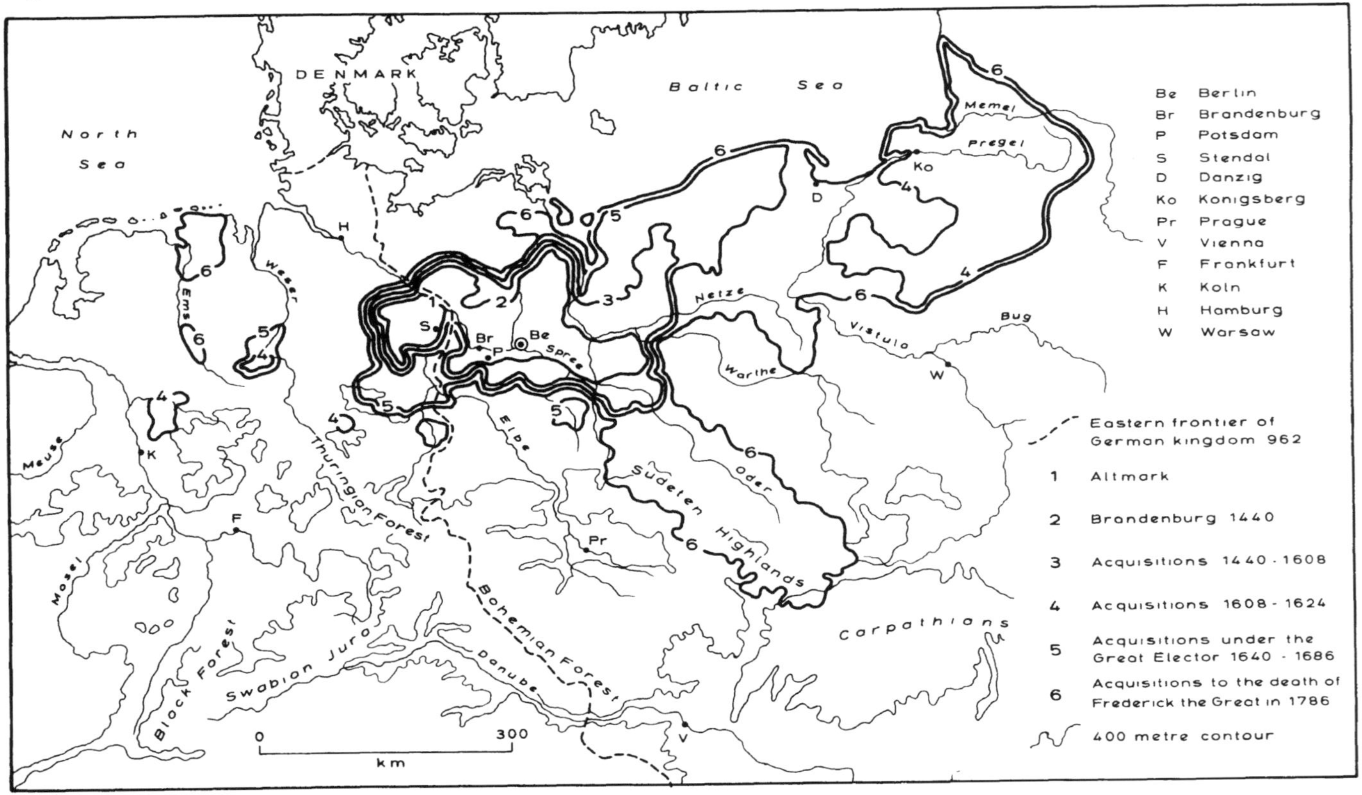

North Sea
DENMARK
Baltic Sea
Memel
Pregel
Ko
D
Ems
Weser
H
Netze
Vistula
Bug
W
S
Br
P
Be
Spree
Warthe
Elbe
Oder
K
F
Meuse
Thuringian Forest
Sudeten Highlands
Pr
Mosel
Bohemian Forest
Carpathians
Black Forest
Swabian Jura
Danube
V
0
300
km
Be Berlin
Br Brandenburg
P Potsdam
S Stendal
D Danzig
Ko Konigsberg
Pr Prague
V Vienna
F Frankfurt
K Koln
H Hamburg
W Warsaw
Eastern frontier of German kingdom 962
1 Altmark
2 Brandenburg 1440
3 Acquisitions 1440 - 1608
4 Acquisitions 1608 - 1624
5 Acquisitions under the Great Elector 1640 - 1686
6 Acquisitions to the death of Frederick the Great in 1786
400 metre contour

maritime and continental peripheries of Europe over the French. At the Congress of Vienna a substantial part of Prussia's recently acquired Polish territory was detached and given to the victorious Russians. By way of compensation Prussia was awarded territory in the German Middle Rhinelands from the Netherlands as far as the rift valley. The slamming of the eastern door on Prussia had the effect of reorientating her towards the west, thus making her more German. During the nineteenth century Prussia built on these new acquisitions in the west and gained effective control over the whole of the North European Plain from the Netherlands to Russia. Her territory also extended deep into the basins of the Rhine, Weser and Oder. The Zollverein, the customs union established in 1834, linked Prussia economically with the south German states and paved the way for unification under her leadership. Three short victorious wars completed the process. The first with Denmark in 1864 detached the provinces of Schleswig and Holstein from the Danish monarchy and brought them into the German Confederation. The second, the Brüderkrieg with Austria in 1866, resulted in the incorporation of both Schleswig and Holstein into Prussia and the exclusion of Austria from the subsequent unification process. Finally the defeat of France in 1870 led to the proclamation of the Deutsches Reich in 1871 together with the acquisition of the provinces of Alsace and eastern Lorraine. The second Reich, as then established, was the result of the victory of the Bismarckian policy of Kleindeutschland over the wider objectives of Grossdeutschland which would have had to take into account the historic position of Austria in German affairs.

The two factors underlying the nature of this new German state were Prussian expansionism and German nationalism, and both were to influence its conduct over the next three quarters of a century. As has been observed, Prussian history until the nineteenth century had been one of near continuous territorial expansion eastwards from its historic core west of the Elbe. Originally this had been part of the great easterly movement of conquest which ground to a halt in the fifteenth century as a result of the stiffening of Slav opposition. The continuing expansion of the autonomous Prussian state, by then largely freed from Imperial constraints, was motivated by the twin requirements of resources and security. The physical resources were acquired both by systematic internal colonisation and by the acquisition of richer territory with movement southward into the Börde zone and the Mittelgebirge. The löss soils of the former were the basis for a productive agriculture able to support a large population, and the northern fringes of the latter were rich in the mineral resources required by industry. Silesia and the Rhinelands together made the most important contribution to increasing the economic strength of Prussia, and the importance of this territory sandwiched as it was between the northern plains and the southern mountains was increased in the nineteenth century by the exploitation of the large reserves of coal, iron ore and limestone found in its carboniferous strata.

The other prime requirement was the security of the Prussian state, forced throughout its history to live with that fateful 'conditio Preussin', the curse of its vulnerable central location. This was Prussia's most intractable geopolitical problem and it was tackled both by the attempt to secure forward defensive frontiers and by the development and maintenance of a large and well-trained army. In accordance with the premise that 'Prussia's frontiers are her armies', the military budget was always a large one and considerable importance was accorded to the military caste. This caste came mainly from the Junkers, the Prussian landowners, whose origins on the eastern marchlands of the Holy Roman Empire fitted them well to become the warrior aristocracy of a militaristic state. Prussia thus often gave the appearance of being an army with a state rather than a state with an army, and the situation developed in which military power was not simply an instrument of policy but a goal in itself.[24]

The second factor underlying the nature of the German state was nationalism and this, in its modern sense, had its origins in the French Revolution; after 1815 young Germans sought to emulate the French and the English by establishing their own nation state. This German nationalism took root during the first half of the nineteenth century, but, despite the efforts of the premature Frankfurt Parliament in 1848, unification remained elusive. The process of the transformation of this German nation into a state was hindered by two intractable geographical problems. The first was the wide dispersal of German speech throughout the whole of central and eastern Europe and the consequent difficulty of creating a clearly defined nation state possessing its own 'Naturgrenzen', or 'limites naturelles' on the French model. The second was the centrifugal character of the national territory with real central authority having been notable by its absence. Political geographers have attempted to identify proto-core regions, but these were usually the political centres of individual German states rather than kernels of Pan-German unity.[25] The physically diverse Mittelgebirge, together with its northern fringes, was divided politically into a number of states which had become virtually independent within the overall loose structure of the Holy Roman Empire. In the event, the creation of a unified German state had to await the evolution of Prussian policy. The three victorious wars confirmed her primacy in Mitteleuropa and her dominance over Kleindeutschland. This latter consisted mainly of the German-speaking areas of the middle Rhinelands together with the upper basin of the Danube. The federal structure of the new Reich was thus built upon its centrifugal past and the deep local loyalties which had evolved over the centuries. It was in the states of the Mittelgebirge that the spirit of independence remained most in evidence, while those of the North European Plain, with a few notable exceptions, were incorporated into a bloated Prussia.[26] From the outset the latter dominated the new Empire. Its King became the German Kaiser and its capital the imperial seat of government. From its ruling class was drawn the Empire's military leadership, and its civil service became the model for that of the federal government. Its historic core region in Brandenburg-Prussia became the political core of the Empire and its eastern frontiers in Prussia came to be endowed with a mythical quality as the German outpost against Slavdom and a reminder of past Teutonic triumphs.

The one crucial area of the national life in which the Prussians were not dominant was the economy. The principal centres of commercial and industrial power lay in the Börde zone and in the Rhinelands. At the junction of the two was the Ruhr, the largest and richest coalfield in western Europe which, as William Manchester put it, 'spoke of Teutonic power'.[27] Within a generation the Ruhr-Rhine industrial complex, 'the anvil of the Reich', was to make the German Empire the most formidable economic power in Europe. By that time the resources of other parts of Germany were also being opened up and Berlin had become a large industrial city with a population of one million. Between 1871 and 1900 the population of the Reich grew from 40 to 60 millions, faster than that of any of its rivals. While in 1800 the populations of the territories of France and Germany had been about equal, by 1900 that of Germany had grown to become half as great again as that of its neighbour. It was this combination of geology and human ecology grafted onto the Prussian military tradition which gave Germany both the strength and the motivation to wage two world wars, and to come close to winning them both. Germany's only consistent partner during her bid for dominance was Austria, the former hegemonial power in Mitteleuropa. Following Austria's exclusion from Kleindeutschland, the new German Empire saw fit to embark on a policy of friendship with her former rival and this led to close collaboration both before and during World War I.

In accordance with Clauswitz's doctrine that attack is the best form of defence, in 1914 Germany attacked on both western and eastern fronts, and in so doing invoked that dreaded 'war on two fronts' which for generations had been the nightmare of German military strategists. The aim, through the implementation of the Schlieffen Plan, was to gain a rapid victory over the French so that attention could then be turned to the east. As a result of the strength of French resistance aided by support from Britain, this rapid victory failed to materialise. Germany was nevertheless able to hold a forward front from Flanders to the Vosges, penetrating deep into the Paris basin. Likewise in the east, with the help of her ally, Austria-Hungary, she was able to hold a front stretching from the Gulf of Riga to the Black Sea. This military success was the main reason for the final collapse of the Russian Empire, and the short-lived Treaty of Brest-Litovsk, signed in 1918 with the new Soviet régime, would, had it been implemented, have given Germany an over-whelming dominance throughout eastern Europe. The threat posed by German military success remained acute until November 1918 when the Reich collapsed in military and economic exhaustion. In the Treaty of Versailles of 1919 the lower Vistula and Oder basins were detached from Germany and transferred back to Poland. The corridor to the Baltic Sea provided for the new Polish state meant that East Prussia was detached from Germany, thus returning to that historical physical separation of Brandenburg from Prussia which had not been ended until the First Partition of Poland in 1772. In the west Alsace-Lorraine was returned to France and the latter was given economic control of the Saar with its valuable coalfield. Most humiliating of all, the Rhinelands were demilitarised and allied troops were stationed on the west bank of the river. The formidable Kaiserreich was transformed into the fledgling Weimar Republic, a weak and truncated state stripped of its eastern sphere of influence and facing massive economic problems.

Out of the political and economic chaos which threatened to tear the country apart emerged the National Socialist (Nazi) party which proposed ultra-nationalist solutions to the country's predicament. Within a decade they had been responsible for the establishment of a Third Reich which immediately embarked on a policy of overturning the Versailles settlement and followed this up by a policy of massive expansion in Europe. The territorial gains of the early years of World War II were more spectacular than those of World War I. They were preceded by a considerable increase in the size of Germany itself, beginning with the return of the Saar, then Anschluss with Austria and, following the four-power Munich agreement, the incorporation of the Sudetenland. This was shortly followed by the dismemberment of Czechoslovakia and the incorporation of Bohemia and Moravia into the new Reich. With the disappearance of both Austria and Czechoslovakia, Kleindeutschland was enlarged at last into a new and formidable Grossdeutschland. Germany, claiming provocation by Poland, then opened World War II by thrusting eastwards and the 'fourth Partition' of Poland, agreed with the Soviet Union, again removed that unfortunately located country from the map of Europe. This was followed by a German Blitzkrieg north into Scandinavia, west into the Low Countries and France and south into the Balkans. By 1941 the apparently invincible Wehrmacht had achieved domination over most of trunk Europe from the Atlantic to the Baltic-Black Sea isthmus (Figure 3.4). The most spectacular victory was that over France in 1940 when, for the second time within a century, French resistance crumbled in the face of German might. The whole of the north of the country was occupied and in the south the residual Etat Francaise, under the Vichy regime, was reduced to satellite status.

It was at this point in the war that Germany decided to attack the Soviet Union with whom she had signed a non-aggression pact a mere eighteen months before. The success of this 'Operation Barbarossa' was at first so total that the Germans penetrated to the outskirts of both Leningrad and Moscow and held an advanced line from there stretching east-wards 800 kilometres to the Volga and then southwards to the foothills of the Caucasus. By the summer of 1942 Germany had conquered the three Baltic republics, Byelorussia and the Ukraine, together with a large part of the peninsula between the Black sea and the Caspian. She was therefore in control of Russia's most important sources of foodstuffs, coal, iron ore and steel, and was approaching the strategically vital oilfields around Baku. However, this proved to be the high water mark of her territorial expansion to the east. Following her catastrophic defeat at Stalingrad in the winter of 1942-43 she was forced into retreat by the steady build-up of Soviet power. At the same time her position in the Mediterranean was being eroded by the build-up of British land and sea power in the region. Her weak Italian ally being soon de-feated, Germany was forced to turn northern Italy into a buffer for the defence of her southern flanks. The Anglo-American air offensive was then followed by a seaborne in-vasion which rapidly opened up a 'second front' in France. Germany came under increasing pressure from all sides, and by the middle of 1945 her bid for supremacy, apparently so effortlessly successful five years earlier, ended in crushing defeat.

The German Empire, as established in 1871, consisted geographically of a thick slice through the principal macro-regions of central Europe. As a result of this, while the frontiers to the north and south had many of the character-istics of Naturgrenzen, those to the west and east lacked any such clear definition. The new state was also located at the very centre of Europe, bounded on its landward side by the continental great powers and with Britain commanding the seas to the north. This physical character and situation were inherited by the Reich from Prussia, and German expansion-ism can be seen as being geopolitically the heir to that of Prussia. However, Prussia was a state and not a nation, and the problem of the creation of a nation-state was something which she had never had to face until she was subsumed into the Reich. The major problem in the creation of a viable nation-state was the inherent indeterminacy of both the

Figure 3.4: Sphere of German Territorial Dominance in 1942

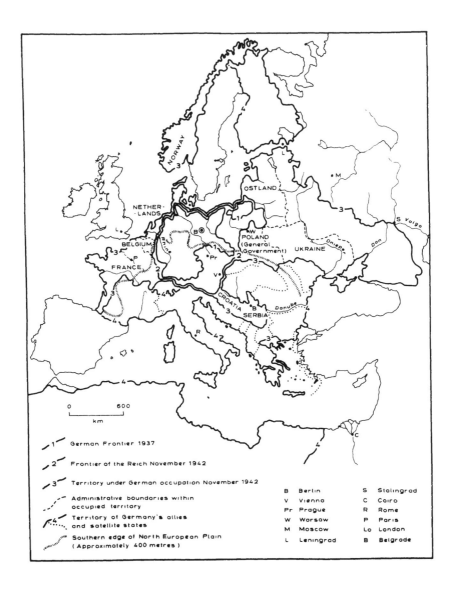

physical and human boundaries. Lacking the relative clarity of France, it could not emulate the latter's confident aspiration towards 'les limites naturelles'.[28] Legitimacy was therefore sought through the unification of the Volk and, in drawing territorial limits, Naturgrenze gave place to Kulturgrenze. This immediately raised the twin problems of the wide dispersal of Germans beyond any conceivably acceptable frontiers and the deliberate exclusion of Austria from Kleindeutschland. The two were closely related since the question of Germany's 'dispersed irredenta' was of major concern to the Pan-German movement in the late nineteenth century, a movement which had a strong following in Austria itself. Another problem was the historical fluidity of the nation's frontiers, especially since the Holy Roman Empire of the German People, at its largest, had covered an enormous part of Europe. Claims were made by Pan-Germans to places as far apart as the Baltic, the Rhinelands and Scandinavia, but any possibility of their realisation was contingent upon Germany's achievement of a position of dominance. A more potentially realisable idea was that of the creation of a central European bloc, Mitteleuropa. This went a long way towards satisfying the Pan-Germans and also those who sought the creation of a German-dominated economic sphere which would serve as both a market for her industrial products and a source of much needed foodstuffs and raw materials.[29] However, here also there was a problem of definition, and the adherents of Mitteleuropa often had very different geographical areas in mind. Partsch envisaged it as a massive area stretching from the Low Countries to the Baltic and from the North Sea to the Alps, a total of one million square kilometres with a population of 130 millions.[30] Its natural outlets were the Rhine and the Danube and many Germans held the opinion that the powers holding the fringes of these rivers, France and Russia, would have to be dislodged from them if Mitteleuropa were to be strong and secure. Most definitions of Mitteleuropa emphasised the importance of eastern Europe, and thus implicitly related it to the ideas of Drang nach Osten and Lebensraum. The former stemmed from the belief, particularly current among 'east Elbians', that Germany's real destiny lay in the east. During the period of the Kaiserreich both Mitteleuropa and Drang nach Osten were implicit in the alliances with Austria-Hungary and the Ottoman Empire, in the cultivation of good relations with the Balkan states and in the development of close economic ties with south-east Europe. A spectacular project which linked economic development and political advantage was the Berlin to Baghdad railway, seen as being a continental alternative to the Suez Canal and outflanking British maritime supremacy. While during World War I Austria-Hungary had still been the hegemonial state in south-east Europe, by World War II the region had been brought directly under German control.

However, on both occasions Germany was enticed into the affairs of the region in support of her flagging allies, Austria and Italy respectively, thus draining badly needed resources from the principal fronts in the west and east.

The conviction that Germany had a particular destiny in the east was also related to the wider geopolitical doctrine of Lebensraum. Originating in the ideas propounded by Ratzel, the father of German political geography,[31] and founded on Darwinist concepts, this was the contention that it was essential for a nation to expand if it were to remain strong and vigorous. The doctrine was given added force by the spectacular growth in the German population to reach nearly 70 millions by the time of World War I. A quarter of a century later it was about the same, but, as a result of the substantial loss of territory in the east, by then it lived on only four-fifths the area. Frequent bitter comparisons were made between Great Britain with a population of 40 millions and a gigantic global empire and Germany with a far larger population but shorn of both colonies overseas and territory in Europe.

Fears arising from the constriction of Germany and the inadequacy of her space had been a feature of German geopolitical thinking since the late nineteenth century, but a persistent dichotomy had existed between the adherents of Weltpolitik and those of Lebensraum.[32] While the formed favoured colonialism, world trade and the construction of a large navy, the latter wanted an extension of German power in Europe and a powerful army. Lebensraum found favour with the Pan-Germans and with the Volk movement which was suspicious of industry and urbanisation and desired the extension of peasant farming, especially in the east. In the conditions prevailing after World War I, it was the Lebensraum doctrine, particularly restoration and improvement of Germany's position in Europe, which prevailed.[33] This soon took the form of a belief that the natural colonial territory for Germany lay in the east, especially in the vast spaces of Russia. After the model of the colonial powers of the west, these territories were to be used to take population surplus, supply foodstuffs and raw materials, and act as markets for German manufactured goods.

In the aftermath of the German defeat in World War I, the study of Geopolitik emerged as a policy-oriented branch of political geography. Its roots went back to the political and intellectual traditions of Prussia and the Second Empire. Its principal practitioners, led by General Karl Haushofer of Munich, sought to employ geography and geographical methodology in the analysis of the interests of the state.[34] They developed theories derived from Ratzel and from the Swedish political scientist Kjellén, who was the first to use the term Geopolitik,[35] into a 'scientific' justification of aggressive and expansionist policies. They countered the physical indeterminacy of Germany with the concept of the Volk as a powerful cultural reality and held strongly to the view that 'the logic of history demands its unification'.[36] Such specific geographical terms as Volksboden, Sprachboden and Kulturboden were developed and theorised over.[37] They were intended to indicate different aspects of the Deutschtum, the area of 'Germanness' in Europe, which at its largest covered an enormous part of the continent.[38] While they basically favoured the Lebensraum idea, they also incorporated many aspects of Weltpolitik into their thinking.

The exact extent to which the Nazi leadership was itself influenced by Geopolitik remains uncertain, but Mein Kampf is filled with semi-digested geopolitical ideas emanating from various sources. The 'Lebensraum imperialism' of the Third Reich was a composite owing something to perceived economic and political needs, geopolitical theories and racialism.[39] The myth of Aryan racial superiority, also founded in social Darwinism, was developed and rationalised by the Nazis. It entailed the necessity to safeguard the Aryan German blood from contamination by those deemed by them to be inferior races, in particular the Slavs and Jews. One of the most sinister features of this particular brand of racism was its fanatic anti-semitism as typified by the ideas of the Führer himself. The origins of this have been seen to lie as much in the multi-national cauldron of pre-World War I Austria as in the racially more homogeneous Reich. Hitler's real historical mission, as he saw it, was the resolution once and for all of the chronic German geopolitical problem.[40] This was to be achieved through Lebensraum and the forced removal of Untermenschen followed by recolonisation with racially and culturally pure German stock.

Although thus strongly influenced by Darwinist spatial and racial ideas and imbued with a particularly ferocious form of twentieth century neo-barbarism, the imperialism of the Third Reich was grounded in historic Prussian and German perceptions of the interests of the state. In particular these included the security of the vulnerable frontiers, the creation of a large economic sphere and the colonisation of new territory. This was to be sustained and justified through an updated version of the German national idea, the Gemeinschaftsgefühl, and the dream of the heroic advance of the superior German people bearing their Kultur and organising the other Europeans through a German-centred hegemony.

While the traditional perceptions of the 'East Elbians' were reinforced by Lebensraum ideas, other Germans were more disposed to see their future in western terms. The Lotharingian axis bound the west of Germany into the communications networks of the adjacent parts of Europe. 'La civilisation rhénane' had been characterised historically by large cities, a flourishing commerce, relative freedom from central authority and close contacts with the Low Countries to the north and Italy to the south. As has been seen, in her hegemonial period France pushed eastwards in search of the completion of 'les limites naturelles' and the final unification of 'la civilisation rhénane with 'la civilisation francaise' seemed then to be a real possibility. For a time the Rhinelands were thus reduced to a satellite status within the French Empire. The German nationalists of the early nineteenth century countered this by asserting that the Rhine was 'Deutschlands Strom aber nicht Deutschlands Grenze'[41] and 'Die Wacht am Rhein' expressed the necessity for constant vigilance against the expansionist enemy in the west.[42] The annexation of Alsace-Lorraine was a move in the direction of Naturgrenzen since the Vosges mountains are the watershed between the Rhine and the Seine-Rhone drainage basins. However, if the geographical logic of this were followed to its conclusion, then the 'Naturgrenze Deutschlands' would have corresponded closely to the western frontier of the Holy Roman Empire, and the Low Countries and a large part of eastern France would have been within it. During both wars claims were made to parts of this territory and it was included in the geopolitical maps as being a part of the Deutschtum.[43] Possession of the whole of the Rhinelands would also afford Germany a major outlet on the southern North Sea, economic control over the Lotharingian axis and a secure 'Vorfeld', a buffer zero around the vulnerable Ruhr-Rhine industrial area. Viewed from the French perspective, this was the culmination of the process of Musspreussen, the forced Prussianisation of the Rhinelands, which began in the early nineteenth century, and its political completion with an 'anschluss rhénane' along the same lines as that Anschluss with Austria.[44]

The leaders of the Third Reich and the practitioners of Geopolitik shared the opinion that the defeat of France was a prerequisite for the establishment of a secure Germany based on a strong Mitteleuropa.[45] Likewise, it was essential to defeat the Soviet Union before the Lebensraum policy could be implemented in the east. There, however, the resemblance between west and east comes to an end. In the west the object was to bring about a frontier rectification based upon what were perceived as being geopolitical realities. In the east on the other hand the object was the achievement of a position of dominance and this entailed the complete reorganisation of that part of Europe. At its most ambitious, it meant the dismemberment of the Soviet Union and the incorporation of its western territories into the sphere. Russia herself was to be reduced to her pre-Petrine frontiers and even cease to exist at all as an independent state.

The territorial expansion of Germany during the last five years of the Third Reich was far greater than that which had taken place during the whole of previous German history, and made it one of the most impressive territorial empires in the whole of modern history. The geopolitical logic behind the acquisition of this enormous empire was present in the three ideas of Pan-Germanism, Mitteleuropa and Lebensraum. These eventually emerged in Nazi geopolitical thinking as arguments for a permanent German domination over the European continent from the tightly drawn frontiers of France on the Vosges to the steppes of Russia 2000 kilometres to the east. This was to be organised in accordance with the ideas of Die Neuordnung, a new order for the whole of the European continent which was to arise upon the ruins of the old Europe. Features of the old order which were to be swept away included the bourgeois-democratic social system, the Christian Churches, the Jews, the Communists and the small and ineffective nation states which were the principal legacy of Versailles. The greatest contempt of all was reserved for this Kleinstaatengerümpel which was to be swept away and replaced by a new National Socialist international system organised and led by Germany.[46] The most feared enemies of 'Die Neuordnung' were the Communists and the Jews, linked together in Nazi propoganda as the 'Jewish-Bolshevik conspiracy', and the German intention was to rally the combined energies of the subject peoples and the satellite states to fight against this spectre which the Nazis had themselves evoked. 'Fall Barbarossa', recalling the glories of the medieval Empire, was presented by Goebbels' propaganda machine as a crusade by the nations of civilised Europe, led by Hitler's latter-day Teutonic Knights, against Asiatic barbarism represented by the Jews and the Communists. 'To the Nazi mind', said Manchester, 'Fall Barbarossa...... was a crusade against evil'.[47]

Given the extent of its ambitions, the defeat of the Thousand Year Reich when it came was all the more cataclysmic. The geopolitical elements in the situation bear many resemblances to those in the defeat of the French Empire. As with the latter, there was a total inability to subdue the maritime and continental peripheries, still held, as they had been a century and a half earlier, by the same two powers. Despite the initial successes of the Blitzkrieg on land, sea and in the air, the advance into distant and alien geographical environments proved to be a total disaster. 'Le général février' and the Red Army in the east and the Mediterranean together with the sea power of the British in the south and west combined to bring the Wehrmacht to a halt on the plains of Russia and in the deserts of North Africa. The battered and beleaguered heartlands of both maritime and continental peripheries held until relief could come from the massive resources behind the Ural mountains and across the Atlantic Ocean on a scale far greater than was available to Germany. Neither could Germany acquire in time sufficient expertise to wage war at the same time either on land against Russia or at sea against Britain when those two powers had centuries of experience of warfare in these environments.

Resources were further sapped by the diversion of German energies southwards into the Balkans and the Mediterranean region in support of the unrealistic ambitions of her Italian ally to dominate Mare Nostro. This was peripheral to Germany's immediate strategic concerns based on Mitteleuropa and Lebensraum, and put her in the debilitating and dangerous position of assisting a weaker power to mount a hegemonial bid over an adjacent geopolitical region.[48] The most fateful consequence of this diversion of energies was the delay in mounting Operation Barbarossa with the result that the Wehrmacht was forced to endure the Russian winter with equipment more appropriate for a summer war in a temperate climate. What had been intended as the ultimate Blitzkrieg dragged on to become the historic nightmare of war on two, and then on three, fronts. This was the ultimate catastrophe resulting from the Conditio Germaniae, a repeat on a massive scale of what had happened to Prussia two centuries earlier. The historical parallel of Frederick the Great was not lost on Hitler, but the hoped-for deliverance failed to materialise. After 1943, with the allies closing in from east, west and south, Grossdeutschland was reduced to a beleaguered 'Festung Europa', vulnerable to the end as a result of its central location and the inadequacy of its natural defences. The Volksboden and Kulturboden, so precisely delineated in the maps of the geopoliticians, were not enough to halt the allied tanks or to turn back their aircraft.

Underlying German military inadequacies was the complete failure of Die Neuordnung to present itself as some great new ideal. Indeed German behaviour in the occupied territories, especially in the east, soon gave the lie to the assertion that these latter-day Teutonic Knights, were the crusaders of a reborn European civilisation, the modern standardbearers of a sort of secular revival of Christendom. The bleak reality was that the other nationalities of Europe, and especially the Slavs, were destined to be no more than hewers of wood and drawers of water for the Herrenvolk.[49] The dawning of this realisation was the principal catalyst for the steady build-up of resistance to Germany both inside and outside the boundaries of her enormous empire. Most important of all, it once more galvanised the combined energies of Europe's two peripheral powers. With the assistance of the Americans, they were able to bring about the Wagnerian Götterdämmerung of the 'Tausendjahrige Reich' after a mere thirteen turbulent and destructive years.

'All empires', said T. Callander, 'are identical in substance'. This, he goes on, is because each of them embodies a principle and each 'reveals itself as a special form of a universal concept based on the plenary authority enjoyed by the Roman holders of the maius imperium symbolised by the fasces'.[50] The processes which have been seen to be at work in those states which attained a position of dominance in the western ecumene have many characteristics in common. An attempt will now be made to identify the geopolitical 'substance' of the dominant state as evidenced in the characteristics of those states examined here.

NOTES AND REFERENCES

1. F. Braudel, *The Mediterranean and the Mediterranean World in the Age of Philip II* (Collins Fontana, London, 1972), p. 621.

2. V. Cornish, *The Great Capitals* (Methuen, London, 1923), p. 143.

3. H.A.L. Fisher, *A History of Europe* (Arnold, London, 1936), p. 338.

4. This term refers to Central Europe in the German geopolitical sense. See Glossary.

5. A.J. Toynbee, *A Study of History* (Royal Institute of International Affairs, London, 1934), vol. 2, p. 180.

6. The Pannonian plain with its huge expanse of flat and low-lying land and its natural vegetation of temperate grassland has been seen as being physically an outlier of Asia west of the Carpathian mountains. It stands in stark contrast to the varied and 'European' landscapes of Austria. Metternich was of the opinion that Asia really began at the Landstrasse on the eastern outskirts of Vienna. Despite their long political association, the Austrians continued to regard the Hungarians as being an eastern people. They saw the Hungarians as having, beneath a thin veneer of civilisation, much of their fierce Magyar ancestors left in them.

7. C. McEvedy and R. Jones, *Atlas of World Population History* (Penguin, Harmondsworth, 1978), p. 90.

8. 'Lotharingia' came into being following the Treaty of Verdun in 843 which divided up the empire of Charlemagne among his three grandsons. Lothair received the Middle Kingdom together with the Imperial title. The part of the kingdom which lay north of the Alps did not prove to be geopolitically viable, partly because of its narrowness and partly on account of the indefensibility of its frontiers. It ceased to exist after the Treaty of Mersen in 870 when its territory was partitioned between its western and eastern neighbours. See also Chapter 2, reference 17.

9. J.F. Gravier, *Paris et le Désert Francais* (Flammarion, Paris, 1947). Gravier contended that 'Paris a devoré les provinces' and thus added to its power and its capacity to sustain its extraordinary position of dominance over the rest of France.

10. W. Kolarz, *Myths and Realities in Eastern Europe* (Drummond, London, 1946), pp. 77-8.

11. J. Ancel, *La Géographie des Frontières* (Gallimard, Paris, 1938), p. 68.

12. J. F. von Bielfeld, *Institutions Politiques* (1760) quoted in T. Hunczac (ed.), *Russian Imperialism from Ivan the Great to the Revolution* (Rutgers University Press, New Brunswick, New Jersey, 1974), p. 354.

13. J. Ancel, *La Géographie des Frontières*, p. 68.

14. A. Sorel, *L'Europe et la Révolution Francaise* (1897), quoted in N.J.G. Pounds, 'The Origin of the Idea of Natural Frontiers in France', *Annals of the Association of American Geographers*, no. 41, 1951, p. 146.

15. N.J.G. Pounds, 'France and "les limites naturelles" from the Seventeenth to the Twentieth Centuries', *Annals of the Association of American Geographers*, no. 44, 1954, p. 55.

16. J. Ancel, *La Géographie des Frontières*, p.73.

17. N.J.G. Pounds, 'France and "les limites naturelles"', p. 55.

18. C. McEvedy and R. Jones, *Atlas of World Population History*, p. 58.

19. A.J. Toynbee, *A Study of History* abridged by D.C. Somervell (Oxford University Press, 1946), p. 470.

20. A.J. Toynbee, ibid., p. 471.

21. Although known to history as Henry I, he was never actually crowned Emperor. The first of the Saxon line to be crowned Holy Roman Emperor was his son Otto I (936-962).

22. J.A.R. Marriott and C.G. Robertson, *The Evolution of Prussia* (Oxford University Press, 1915), p. 37.

23. The king of the swamps or borderlands.

24. E. L. Jones, *The European Miracle* (Cambridge University Press, 1981), p. 135.

25. M.I. Glassner and H.J. de Blij, *Systematic Political Geography*, 3rd. ed. (John Wiley, New York, 1980), pp. 94ff.

26. The small states of Mecklenburg and Oldenburg together with the free cities of Bremen and Hamburg were all that remained in the northern plain by the time of the proclamation of the German Empire.

27. W. Manchester, *The Arms of Krupp* (Bantam Books, London, 1970), p. 689.

28. N.J.G. Pounds, 'The Origins of the Idea of Natural Frontiers in France', pp. 150 ff.

29. F. Naumann, *Central Europe* (P.S. King, London, 1916).

30. J. Partsch, *Central Europe* (Heinemann, London, 1903).

31. F. Ratzel, *Politische Geographie* (Oldenburg, Munich, 1897).

32. W.D. Smith, *The Ideological Origins of Nazi Imperialism* (Oxford University Press, 1986), Chapter 6.

33. W.D. Smith, ibid., Chapter 9.

34. G. Parker, *Western Geopolitical Thought in the Twentieth Century* (Croom Helm, London, 1985), Chapter 5.

35. R. Kjellen, *Staten som Lifsform* (Stockholm, 1916) translated into German as *Der Staat als Lebensform* (Hirsel, Leipzig, 1917).

36. J.H. Patterson, 'German Geopolitics Reassessed', *Political Geography Quarterly*, vol. 6, no. 2, April 1987, p. 112.

37. R.E. Dickinson, *The German Lebensraum* (Penguin, Harmondsworth, 1943).

38. W.D. Smith, *The Ideological Origins of Nazi Imperialism*, p. 69.

39. M. Bassin, 'Race contra space: the conflict between German *Geopolitik* and National Socialism', *Political Geography Quarterly*, vol. 6, no. 2, April 1987, p. 115.

40. D. Calleo, *The German Problem Reconsidered* (Cambridge University Press, 1984).

41. 'Germany's river but not Germany's frontier'. From a poem by J. Arndt, written in 1813.

42. Lieb Vaterland, Kannst ruhig sein,
    Fest steht und treu sie Wacht,
    Die Wacht am Rhein.
From a poem by Becker, written in 1853.

43. An example is in Karl Haushofer's *Raumüberwindende Mächte* (Teubner, Leipzig, 1934) reproduced in G. Parker, *Western Geopolitical Thought*, fig. 5.4, p. 64.

44. J. Ancel, *La Géographie des Frontières*, p. 185.

45. G. Parker, *Western Geopolitical Thought*, Chapter 6.

46. A 'rubbish of small states'. The term was used contemptuously by Joseph Goebbels when referring to the old order in Europe and the need to sweep it away and replace it with the Nazi *Neuordnung*. See *The Goebbels Diaries. The Last Days* ed. H. Trevor-Roper (Secker and Warburg, London, 1978).

47. W. Manchester, *The Arms of Krupp*, p. 480.

48. The *Drang nach dem Süden* had been characteristic of the Holy Roman Empire with its close cisalpine connections. The Prussian-dominated *Kleindeutschland* of the late nineteenth century was mainly interested in northern European affairs and Teutonic Europe north of the Alpine-Carpathian divide was certainly the principal preoccupation of the Third Reich.

49. G. Parker, *Western Geopolitical Thought*, p. 81.

50. T. Callander, *The Athenian Empire and the British* (Weidenfeld and Nicolson, London, 1961), p. 11.

FOUR

A GEOPOLITICAL MODEL OF DOMINANCE

As a result of the unique combination of environmental circumstances in which each has operated, the patterns of spatial behaviour in the five states under examination are extremely complex. Nevertheless, recurring spatial patterns can be discerned, and from these it is possible to isolate common spatial characteristics. A list has been drawn up of those geopolitical characteristics which are common to at least two states (Table 4.1). Of the 32 characteristics identified in this way, 11 of them are present in all five states and a further ten are found in four of them. This means that 21 of the characteristics are present in at least four of the five states, a total incidence of 78 per cent. These characteristics will now be identified and will then be used in the construction of a normative geopolitical model of the dominant state. The characteristics are identified chronologically, and the model which is constructed from them is a dynamic spatial one.

The initial geopolitical characteristic of the dominant state is that its original core area is located at the junction between the territory of its parent culture and that of a different culture. The original cores of Castile and Austria lay at the boundary between Christendom and Islam, while that of the Ottoman Empire lay between Islam and Byzantium. Likewise the Altmark lay on the frontier between the Holy Roman Empire and the heathen Slavs. The core was established within the marchlands around the peripheries of the parent culture for the dual purpose of providing protection and of being a base for further territorial expansion. The core then expanded into the territory of the other culture, tending to gravitate towards a centre from which territorial power could most effectively be wielded. This was usually a centre of hydrographic convergence or divergence with good communications facilitating political control and further expansion. Thus the Ottomans moved to the Sea of Marmara, the Castilians down the westward flowing rivers from the Meseta to the Atlantic, the Austrians into the Pannonian basin and the Prussians eastwards along the urstromtal to the Oder. While the Austrians succeeded in achieving dominion over middle Danubia, their actual control was limited by the fact that the Magyars, nominally their subjects, remained firmly in possession of the hydrographic centre.

France is the only apparent exception to this early pattern of expansion, since the core of the French state was already located at the centre of the richest and most easily controlled area in the country. France would thus appear at first sight to have jumped the earlier stages and established the principal centre of power in the country's major hydrographic basin. However, the early Frankish lands had been to the east of the frontier of Roman Gaul, and expansion took place westwards from there into the Paris basin. Thus France did not so much jump the initial stages of state development as go through them at a much earlier period. The original core and its hydrographic extension then went on to develop into the Frankish Empire which successfully, albeit briefly, united the basins of the Seine and the Rhine. Thus a similar process can be identified to that which took place in the other four states, but occurring over a much longer span of time. The Frankish Empire, which reached its greatest territorial extent under Charlemagne, can therefore be viewed geopolitically in two ways. From one perspective it is a prototype European universal state, preceding the five which have been examined but possessing similar geopolitical characteristics. From another it can be viewed as being an early and short-lived expansion of the first French territorial state. This proved to be highly precocious, and it was nearly a thousand years before France again secured a dominant position.

Returning to the model, the expansion of the historic core into the hydrographic power centre was followed by the formation of a core nation with a high level of linguistic and cultural homogeneity. This nation was located astride the original and the conquered territories, and homogeneity between the two sections was reinforced both by migration and by centralising state policies. Austria did not fully conform with this pattern since the Austrians were part of the wider Deutschtum and did not become a clearly defined nation. The Prussians subsequently overcame this same core nation problem through the implementation of the Kleindeutschland policy. In view of their exclusion from the unification of Germany and the multi-national character of their dominions, this was not an option readily open to the Austrians. From the beginning they were weakened by their inability to consolidate their control over the central hydrographic basin. The core nation was subjected to strong pressures emanating from both inside and outside the culture area, and this was followed by counteractive territorial

Table 1: Geopolitical Characteristics of the Dominant State

	Ottoman Empire	Spain	Austria	France	Germany	Total
* 1. Original core of state on frontier of parent culture	x	x	x	(x)	x	4
* 2. Original and historic core in relatively poor area	x	x	x		x	4
* 3. Historic core in interior location	x	x	x	x	x	5
4. Acquisition of coastline by historic core state	x				x	2
* 5. Historic core less developed than parent state	x	x	x	(x)	x	4
* 6. Historic core state centralised, absolutist and militaristic	x	x	x	x	x	5
7. Original and historic core marches of parent state but gained independence	x		x		x	3
* 8. Gravitation towards centre from which control easiest to exercise	x	x	x	x	x	5
* 9. Early acquisition of more valuable territory	x	x	x	(x)	x	4
10. Subsequent acquisition of more valuable territory which becomes principal economic centre		x	x		x	3
*11. Capital moved to forward location within conquered territory	x	x	x	(x)	x	4
*12. Capital remains in historic core (in original or conquered territory)		x	x	x	x	4
*13. Historic core detached and remote from macrocore of culture	x	x	x	x	x	5
*14. Historic core feudal-agrarian	x	x	x	x	x	5
15. Colonisation of conquered territory from core nation	x	x			x	3
*16. Ethnic and cultural diversity around frontiers of expanded core	x	x	x		x	4

x Characteristic present; * Characteristic used in model; (x) French anomaly (see text)

Table 1: (continued)

	Ottoman Empire	Spain	Austria	France	Germany	Total
*17. Imposition of uniformity throughout core state	x	x	x	x	x	5
*18. Expansion triggered by external pressures and events	x	x	x	x	x	5
19. Social/political upheaval in core nation precedes bid for domination				x	x	2
20. Expansion encouraged by unrest and power vacuum on frontiers	x	x	x			3
21. Expansion to attain 'natural' frontiers			x	x	x	3
*22. Control over macrocore of culture as successor to parent state	x	x	x	x	x	5
23. Expansion into uninterrupted contiguous lowlands			x		x	2
*24. Expansion into regions with similar physical conditions	x	x	x	x	x	5
*25. Homogeneous population group becomes core nation	x	x		x	x	4
*26. Bid for dominance in two stages viz (a) regional hegemony (b) universal state	x	x	x	x	x	5
27. Stage (b) durable (min. two generations)	x	x	x			3
28. Control over 'sacred places' of the culture	x	x		x		3
*29. Imposition of order throughout territory controlled	x	x		x	x	4
*30. Imposition of political, cultural and economic uniformity	x	x		x	x	4
*31. Geographical 'logic of unity'	x	x	x	x	x	5
32. Attempt to create a closed system in area of control		x		x	x	3
Totals	26	27	24	19(23)	29	125(129)
Total %	81	84	75	59(72)	91	78(81)
Total characteristics used in model	21	21	18	15(19)	21	96(100)
Percentage conformity with model	100	100	86	71(90)	100	91(95)

x Characteristic present; * Characteristic used in model; (x) French anomaly (see text) expansion by the core state which in this way considerably enlarged the area under its control.

In all five states this led to an attempt to secure control over the macrocore of the culture area. This having been achieved, an attempt was made to impose uniformity first upon it, and then upon the remainder of the culture area. This was done in the name of an ideological world-view normally having its origins in that of the macrocore itself. This was then reorientated to focus on the pivotal role of the core nation in the perpetuation and dissemination of the world-view. The particular world-views espoused by the dominant states under examination are Islam by the Ottoman Empire, Catholicism by Spain, the Enlightenment by France and Social Darwinism by Germany. The exception to this was Austria which, after the early crusading period, was sustained principally by allegiance to the Habsburg dynasty. Dynastic allegiance as a substitute for ideology became increasingly unsatisfactory when it came to be confronted by the powerful new ideology of nationalism which emerged as a major disintegrative force in the nineteenth century.

Finally there was the concurrence of political and geographical elements which produced a 'logic of unity' within what was perceived to be in both physical and human terms a 'natural' unit of organisation. In the case of both the Ottoman Empire and Spain this was the Mediterranean, over the eastern and western basins of which they asserted their respective hegemonies while both laying claim to the totality of the inheritance of the Imperium Romanum. In similar fashion Austria aspired to dominance over the Danubian basin, France to the Rhinelands and Germany to the North European Plain from the Vosges to the Russian steppes. Thus geography both underlay and helped sustain the ideologies of dominance by supplying both the territory and the justifications for expansion into it.

Characteristics which have a slightly lower overall incidence, that is to say those which are found in only three of the states, include that of the original core having been established by, and having remained a part of, the imperial state of the parent culture. This was certainly so with the original Osmanli territory, the Altmark and the Ostmark, but it was not a feature of either the Frankish core or Castile. The latter was an autochthonous state and, unlike Aragon, its territory was never within Charlemagne's buffer zone of the Marca Hispanica. The pre-hegemonial acquisition of a powerful economic base characterises Spain's control of the Lotharingian axis, Prussia's acquisition of the middle Rhinelands and, less successfully, Austria's inheritance of the Netherlands. However, such a development is less evident in France or the Ottoman Empire, although the Balkans did for a time constitute a sort of 'treasure house' in Braudel's sense. Again, the successful and purposeful expansion of the core state towards the sea is a characteristic only of Prussia and the Ottoman Empire. The attempt to secure 'natural' frontiers was also by no means general. While it was a central feature of French policy, and very important to both Austria and Prussia, it is not much in evidence in either the Ottoman Empire or Spain, both of which laid considerable emphasis on maritime conquest and dominance. Expansion into contiguous lowlands is, however, a feature of German, Austrian and Ottoman dominance. Finally, the dominant state proved to be very durable in the case of the Ottoman Empire and fairly so in those of Spain and Austria. In the cases of both France and Germany, however, it proved to be something of very brief duration, and the preceding hegemonial phases lasted far longer.

Both the Ottoman Empire and France have certain features atypical of the process of domination. In the former this was the occupation of the core region of the Byzantine Empire and the transfer there of their capital city and major centre of power. The Ottomans therefore became the successor state to the defunct empire, their former imperial rival, before they had succeeded in achieving a position of dominance with their own culture area. Also atypical are certain features of France, even allowing for the anomalies which have already been noted. Following the failure of the precocious expansion associated with Charlemagne, France came to consititute a semi-autonomous sub-culture embedded within the loose overall framework of Christendom. Gradually it gained many of the characteristics of an alternative culture possessing an alternative macrocore of its own. This underlay the hostility of successive Popes and Emperors towards France, and the attempts to isolate her and keep her frontiers tightly drawn. Within the French macrocore centring on Paris the new secular ideology of the Enlightenment began to evolve in the seventeenth century, and in the eighteenth it was to mount a major challenge to the 'ancien régimes', the historic power structures of Europe. This French alternative culture area neither encouraged nor even permitted the considerable political devolution which characterised the German lands from which Austria and Prussia had emerged. Political power was from the outset consolidated in Paris, and the development of a centralised state was inimical to peripheral autonomy. Strasbourg remained a fortress and a symbol and showed no signs of becoming an autonomous centre of power after the manner of Berlin or Vienna.

Thus the French expansionist phase was generated from a rich centre rather than from a poor periphery as in all the other cases. However, on closer inspection, French behaviour patterns are again not so atypical as might at first appear. In pre-industrial Europe the Paris basin has been one of the wealthiest parts of the continent. By the late seventeenth century this favoured region was already beginning to fall behind other parts of northern Europe which were proving to be more congenial locations for the new developments in commerce and industry. The first phase of French expansion can thus be seen as having been in part an attempt to rectify an increasingly unfavourable economic position through territorial aggrandisement. While in the other four powers it was in the isolation and comparative poverty of the periphery that expansionism was generated, in France it was the relative decline of the country's core region, leading eventually to a massive political and economic collapse at the centre, which produced a very similar effect.

The incidence of all the characteristics in Table 4.1 will now be summarised. As was observed at the outset, the total incidence throughout the five states is 66 per cent. However, this ranges from Germany which has 90 per cent of them down to France with just under 60 per cent. If the apparently anomalous features of the French core region are interpreted as being a more protracted process of development rather than a true anomaly, then France rises to 72 per cent, which is very close to the incidence found in examining Austria. Spain and the Ottoman Empire have 84 and 81 per cent respectively. It is only the 21 particular characteristics found in at least four of the states which will be used in constructing the model. The Ottoman Empire, Spain and Germany possess every one of these, while Austria has an 86 per cent incidence of them. Once again, if the apparent French anomaly is allowed for, then that of France becomes very close to that of Austria.

The model of the dominant state is a dynamic spatial one consisting of three stages from the origins of the state to its achievement of a dominant position. In Stage 1 the original core is located at the junction of the parent culture (A) with the other culture (B) (Figure 4.1). The junction is marked by a fluid frontier zone in which there is considerable movement. The core was established by the authority of culture A for the purpose of bringing stability to the region and of turning the indeterminate situation within it to its own advantage. This is effected by the dominance of a military-religious elite embarked on conquest and conversion. The state is physically poor with a limited and basically agrarian economy. It, together with the whole of the periphery of which it constitutes a part, has none of the wealth, variety and cultural sophistication of the macrocore. This latter is in the centre of the culture area at the focus of its natural routeways, and the most important political, cultural, commercial and industrial centres are located within it. Its great diversity engenders a relatively tolerant intellectual climate. In contrast crude, simplistic and fundamentalist attitudes are characteristic of the periphery. Juxtaposed as it is to an alien culture, these are transformed into a hard and unyielding ethic which pervades the emerging core state. It evolves a centralised and militaristic political structure,

Figure 4.1: Geopolitical Model of Dominance: Stage 1

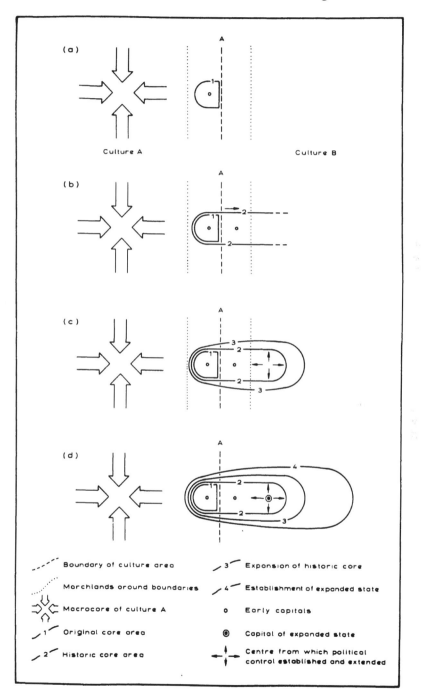

geared to war and aggression in the name of what it inter-prets to be the fundamental values of its culture. The vigour of the original core manifests itself in territorial expansion, and the frontier becomes a dynamic one, moving forward into the territory of culture B. After the initial expansion, move-ment gravitates towards an area from which political control can be most effectively exercised. This is normally a con-vergence or divergence of rivers or a natural routeway which is both a centre of communication and a generator of the resources required by the emerging state. The acquisition of new resources in the territories conquered adds substantially to the strength of the state and thus to its capacity for further expansion. The leadership of the core now also con-stitutes the dominant elite of the conquered territories, and it imposes its own religion and political system upon them. Gradually the core sheds the characteristics of a frontier mark and becomes a power in its own right, no longer under the direction either of the parent state or of the central political authority of the culture. The original core itself, and even some of its early acquisitions, remain within the overall quasi-imperial framework of the culture, but the conquered lands are increasingly treated as being a private preserve. The expanded core state orients itself towards them, and their further extension, consolidation and economic develop-ment becomes a major priority. The persistence of conflict on the forward frontiers with the much depleted but still signifi-cant forces of culture B has the effect of keeping the core state on a dynamic military footing and promotes the conquest and absorption of further territory.

Stage 2 is reached when the core state swings around to reorient itself back into the territory of its parent culture (Figure 4.2). The spatial expression of this is territorial acquisition within the culture area, culminating in the achievement of a dominating position within the macrocore itself. In so doing it takes on the role of protector state of its own parent culture. Physical possession of its 'sacred places' - both secular and religious - strengthens its claim to be the successor state to the former political authority. The relatively high levels of cultural homogeneity which are characteristic of the central parts of the parent culture constitute the justification for the initiation of measures to achieve greater uniformity within all the areas under the control of the new core state. In this way a 'core-nation' now emerges possessing high levels of cultural uniformity, includ-ing language, as well as an overall acceptance of being a unit for political purposes. This unit is located astride the old frontier, partly within the parent culture and partly in the conquered lands. Centrifugal structures formerly character-istic of the culture are replaced by centripetal ones. The increased centralisation thus made possible leads to the eros-ion of the power of regional centres and to a dimunition in

Figure 4.2: Geopolitical Model of Dominance: Stage 2

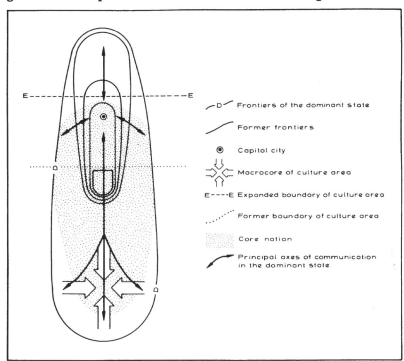

their cultural, political and economic significance. The new situation is reflected in a removal of internal barriers and, at the same time, a reinforcement of external ones, the reformulation of internal political boundaries and the development of networks of communication reflecting the pre-eminence of the capital city located within the historic core region. The characteristics of the core nation are made up of a fusion of those of the culture as a whole with those specific to the core state itself. Foremost among the latter is a continuing propensity towards territorial expansion, arising out of the continued role of the military caste. This provides the impetus for a forward movement towards what in some instances are considered as being 'natural' frontiers - those which are judged necessary to the security and proper functioning of the state. With the completion of Stage 2 the expanding state has successfully united two dissimilar geopolitical regions, the one lying within the historical territory of the parent culture, while the other consists of territory gained from the opposing culture. The former remains relatively rich and diverse, while the latter is poorer, more dependent and retains residual elements of its former culture. The characteristics of the core nation itself are close to those of the parent culture within

Figure 4.3: Geopolitical Model of Dominance: Stage 3

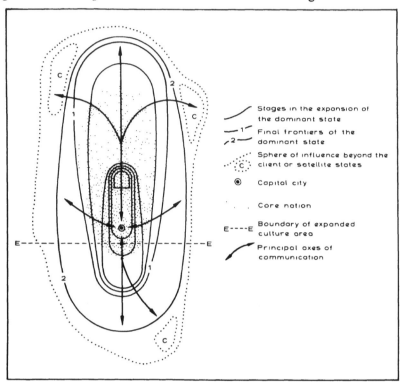

the latter's historic territories, but are closer to that of the core state near and across its former frontiers.

Stage 3 is reached when a position of dominance has been attained. It involves the assumption of control over most of the territory of the parent culture and the unification of this with a large part of the former territory of culture B into an overall political framework. The peripheries into which the dominant state now moves diverge considerably from the norms of the parent culture, and semi-autonomous sub-cultures may have risen within them. As a result their absorption presents considerable difficulty, and requires the expenditure of large resources. Consequently a more practical option on the peripheries is the establishment of satellite states, and a ring of such states may form an extended buffer zone around the territory directly controlled by the dominant power. They may also serve as part of a cordon sanitaire between the dominant state of one culture and that of another. The dominant state then imposes those elements of uniformity which have for long been characteristic of its own core state into all parts of its dominions, and even into the territory of the satellite states themselves. The imposition of uniformity has the practical advantage of easing the exercise of supreme power from the centre, but more important it is also a necessary part of the process of imposing that ideological purity which has been a major justification of dominance. Such purity is the outcome of the strict and literal interpretation of the particular world-view espoused by the dominant state and universalised through its control over the ecumene. It springs from the 'soul' of the core nation, as interpreted and guided by its spiritual or secular high priests, and then serves as the principal weapon in the extension of the authority of the state. Its revealed truths are presented as being so self-evident that any oppositon to them, and to the advance of the dominant state which encapsulates them, must therefore be the result of evil or misguided intentions. Since this view is unlikely to be accepted automatically by the whole heterogeneous population of the ecumene, uniformity will need to be imposed by the central government and those elements which still refuse to accept will then be subject to coercion.

Dominance is thus expressed geopolitically as the imposition on the whole of the territory of the dominant state of the characteristics of the core state. The uniformity of the internal political structure is designed to make the parts fit more readily into the wider whole. The parts will then be grouped around the core nation which remains central to the working of the system, with its 'sacred places' becoming also those of the territory as a whole. It enshrines the world-view of the dominant state and is the principal motor for its extension and consolidation. The process is completed when monostatism and cultural uniformity finally triumph over polystatism and cultural diversity. This, then, constitutes the geopolitical 'substance' of the dominant state. The model will be used in the interpretation of the emergence of Russia as an expansionist state and the subsequent development of Russian imperialism.

FIVE

FROM OSTROG TO EMPIRE: THE TERRITORIAL
EXPANSION OF RUSSIA

'The history of Moscow', said Kerner, 'is the story of how an insignificant ostrog became the capital of a Eurasian empire'. [1] The scale of its expansion dwarfed anything which had previously taken place in the western ecumene. The stronghold of Moscow was transformed first into the principality of Muscovy, then into Russia and finally into the gigantic Russian Empire. By the eighteenth century this was the largest state in the world and one of the acknowledged great powers of Europe. Together with Great Britain, it was one of the key members of that alliance which defeated the French bid for supremacy in the western ecumene. However, unlike France following her defeat of Spain, Russia did not follow up her victory by herself moving into a dominant position. Despite her tremendous strength, prestige and influence at the end of the Napoleonic Wars, Russia did not secure control over either of the macrocores of the western ecumene or over its principal axes of communication. Nevertheless, its massive size was already ample testimony to that expansionist spirit which had taken the double-headed eagle of the Tsars eastwards to the shores of the Pacific and southwards to the deserts of central Asia. This expansion was founded on a strongly held world-view which constituted its justification and at the same time supplied its objectives. The logical culmination of Russian expansionism was thus the achievement of a dominant position. In attempting to explain why this did not come about, it is necessary first to examine the physical conditions in which the Russian state originated and developed.

Moving eastwards across Europe, the Hercynian and Alpine mountain zones dip southwards and give place to the massive enlargement of the North European Plain which stretches eastwards to the Urals and southwards to the Black Sea and the Caspian. The principal areal differentiation of this otherwise physically homogeneous region is provided by the vegetation belts aligned from west to east across it. A considerable range of vegetational types is to be found in the latitudinal spread from 40° north to the Arctic Circle and beyond. These are scrub and semi-desert, temperate grass-land (Steppe), deciduous woodland, coniferous forest (Taiga) and finally the short grass and moss (Tundra) of the Arctic regions. These vegetational zones are linked by the great rivers which traverse the plain. The largest of them, includ-ing the Volga, Don and Dniepr, flow southwards, but there are also a number which flow northwards to the Arctic Ocean and west to the Baltic. The major watershed is aligned from south-west to north-east and passes through the Valdai Hills, one of the few elevations of over 300 metres. These hills lie within the deciduous and mixed forest belt which forms a triangle with its base in the west and its apex in the southern Urals, and this is the historic homeland of the East Slavs, the population group to which the Russians belong.

The first Russian state, Kievan Russ, had its origins in the river system of this area. Here the early Viking 'men of Russ', supposedly led by their eponymous ancestor Rurik, came in search of wealth from trade or plunder. Before long they had succeeded in establishing a routeway from their Baltic homeland to Constantinople via the Neva, the Volkhov and, by means of portages, the Dniepr. Along this 'route from the Varangians to the Greeks' the Vikings founded towns, the most important of which were Novgorod and Kiev. These became prosperous commercial centres in their own right, handling the growing commercial relations between the capital of the Byzantine Empire and northern Europe. This routeway had the great advantage that it was possible to use water transport most of the way and there were none of the mountain barriers which impeded north-south communication further to the west. As a result of this the principal cities of the Rurikovichi grew in size and wealth. The more northerly ones, in particular Novgorod, gravitated towards the Baltic-North Sea commercial sphere, while the more southerly ones became part of the sphere of the Byzantine Empire. Kiev - 'Byzantium on Dniepr' - outshone all the others and itself became an imperial capital with dependent territories extend-ing from the Dniepr basin to that of the upper Volga. The mixed forest lands which made up this Kievan Russ were bound together physically by the large and comprehensive river systems, and economically by the commerce which was carried on them. When Kiev achieved ascendancy over Novgorod, her territories were expanded northwards into the Taiga and the rich fur trade from there added greatly to her wealth. From south of Kiev to the Black Sea the Dniepr flows through the Steppe, historically the home of nomads from the east. It was from this direction that Kiev was most vulnerable to attack, and the hammer blow came in the thirteenth cen-tury when the Mongols, united under Gengis Khan, surged westwards. The cities of Kievan Russ were sacked, her lands devastated and her empire destroyed. Invincible in the

Eurasian steppes, the Mongols were however unable to master the alien forest environment, and there the residue of the Russians were able to maintain a degree of autonomy as vassal states of the Mongol Empire. At the same time there was a significant population shift northwards from the Mongol held south into the comparative safety of the deeper forests.[2]

Especially attractive to immigration was the Mezhdurechie, the 'Russian Mesopotamia' located between the Volga and the Oka rivers. Here the forest-brown soils produced fertile agricultural land and the rivers provided good communication with the surrounding areas. It was in this region that the new Russian state which was to replace Kievan Russ had its origins. This centred first on Suzdal and Vladimir and then, from the middle of the thirteenth century, on Moscow. The princes of Muscovy first rose to power as the gatherers of taxes for their Mongol overlords, but by the fifteenth century they had themselves become sufficiently powerful to be able to withold this tribute. Ivan III was the first Muscovite prince to claim overlordship of the Russian lands in place of the Mongols and to assume the prestigious title of Tsar. In so doing he was emulating the claim of the Patriarch of the Orthodox Church to spiritual authority over 'all the Russias'. Following the fall of Kiev, the Metropolitan of the Orthodox Church had transferred his seat first to Vladimir and then to Moscow and two centuries later, after the fall of Byzantium to the Ottoman Turks, the Metropolitan had assumed the Patriarchate. By this time Muscovy had evolved into an absolutist, centralised and theocratic state. It was very different in character from the great commercial empire of Kievan Russ, but it bore many resemblances to the Mongol Empire which had dominated it for so long. Muscovy then embarked on a policy of territorial aggrandisement which within a century was to transform it into an empire of massive dimensions extending over the great plain to the Urals and beyond (Figure 5.1).

This expansion took place in a largely concentric manner outwards from the Muscovite core region. The spread to the north was a consequence of the defeat of Novgorod and the acquisition of the enormous sub-Arctic territories to which it claimed title. The second direction of Muscovite expansion was to the south-west into the empire of Poland-Lithuania. Here the Muscovites came into contact with the fierce Cossacks of the lower Dniepr, a mixture of Slav and Mongol, adapted to the harsh life in this okraina, the lawless and indeterminate frontier zone between the dominions of the Lithuanians and the Tartars. As a result of victory here, the frontier was moved westwards to the Dniepr, and Kiev itself rejoined the Russian lands, now in the subordinate role of frontier bastion of Muscovy. Thirdly, there was the easterly expansion inaugurated with the capture of Kazan by Ivan IV in 1552 and the subsequent advance down the Volga to Astrakhan and the

Figure 5.1: The Geopolitical Centre of Muscovite Russia

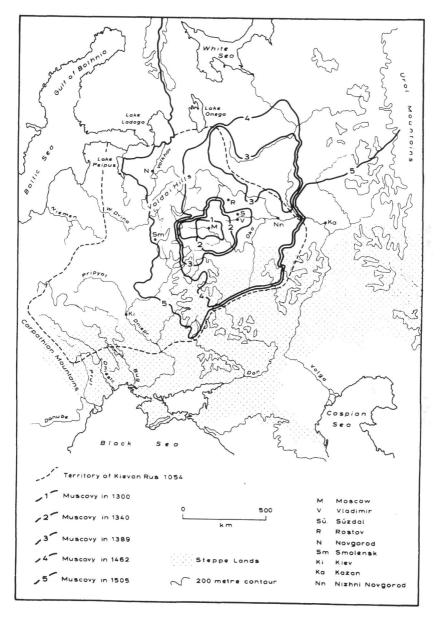

Caspian Sea. This was followed by the annexation of almost the whole of the Don and Volga basins and the opening of a routeway to the east.

Until the sixteenth century the principal theme of Russian political geography had been the conflict between the peoples of the deciduous forest and those of the Steppe and the oscillation of power between them.[3] From this time on the forest peoples captured the initiative and moved successfully to subjugate the steppe-dwellers. The former frontier, a broad belt stretching eastwards from the Dniepr to the southern Urals, was pacified and incorporated into the Russian state; by the nineteenth century it had been transformed into the economic core region of the Empire. In the early years of the eighteenth century the Tsar Peter I had taken the dramatic step of transferring his capital to a new location 600 kilometres north-west of Moscow. Here on the estuary of the river Neva, on land not at that time even recognised as being Russian territory, he began the construction of St. Petersburg. This massive geographical displacement of political power from the centre to the periphery of his domains was a powerful statement of faith in his vision of what Russia could become. To achieve it, landlocked, theocratic and traditionalist Muscovy had to be replaced by a modernised and partially secularised state founded on maritime power. This purpose was inherent in the location, functions and architectural styles of the new capital which was deliberately created to be the antithesis of Moscow.[4]

This internal geopolitical change also inaugurated a new phase in Russian expansion which within a century gave the country improved access to the sea, along with firmer frontiers. At the time Sweden was still the dominant power in northern Europe and she had come to regard the Baltic as being the 'mare internam' of her empire. However, Russia proved victorious in the Northern War and the Treaty of Nystadt in 1721 confirmed her possession of the Baltic provinces of Livonia, Estonia, Ingria and eastern Karelia. Later on in the century, Russia's share of the Partitions of Poland added most of the basins of the Dniepr and western Dvina. In this way she entered fully into her claimed inheritance as successor state to the Kievan Empire. Russian expansion then turned southwards and, following her victories over the Ottomans, she acquired the lower Dniepr, the Crimean peninsula and the northern coast of the Sea of Azov. Her presence on the Black Sea for the first time was marked by the establishment of the commercial port of Odessa and the naval base of Sevastopol. Following the victory over France at the end of the Napoleonic Wars, the Russian frontier was pushed still further to the west with the acquisition of Finland, Bessarabia and most of the Prussian and Austrian shares of the Polish Partitions. The western frontier, saving only for the westerly bulge produced by the acquisition of Poland, now followed closely the 1000 kilometre isthmus connecting the Baltic to the Black Sea, and was consequently shorter and more defensible than at any previous time in the country's history.

Since the sixteenth century conquests the southern isthmus between the Black Sea and the Caspian had remained an unstable frontier zone. The indigenous population, much of it Islamic, was divided in its attitudes to Russian penetration; some welcomed the Russians as liberators while others saw them as being yet another in a succession of alien conquerors. The process of consolidation was sustained by the perceived need for a firm southern frontier and for a long time this followed the line of the Terek river north of the Caucasus. However, at the end of the eighteenth century the Russians were drawn southwards across the mountains in support of Georgia. This ancient Christian kingdom was sandwiched uncomfortably between the Islamic Ottoman and Persian Empires and it had suffered grievously at the hands of them both. As the power of these two empires diminished, Russia moved against them and in 1804 Georgia was finally incorporated into the Russian Empire. In the interests of security on this new southern frontier the Russians moved forward to new positions and by 1830 the whole of Transcaucasia as far as the Aras river had been incorporated within their Empire. The tribes of the Caucasus, now outflanked by the Russians, rebelled and a fierce mountain war, taking on the dimensions of an Islamic Jihad, tied down large Russian forces for many years.

This war did not come to an end until well into the nineteenth century and the Caucasus was thus the last part of the territory of European Russia to be pacified and brought under control. However, by that time dissatisfaction was widespread throughout the Empire, nowhere more so than in Poland which, for cultural and historical reasons, was always the most difficult province to govern. Despite this, the growing dissatisfaction was in general assuming a social rather than a regional or national character, and it was coalescing into revolutionary movements, many of them dedicated to the violent overthrow of the autocracy. The one which proved eventually to be the most important was founded on the potent new ideology of Marxism and during the first quarter of the twentieth century it was to overthrow the Empire and to erect a new revolutionary state in its place.

Just as it is possible to distinguish Castile from Spain and Prussia from Germany, so it is possible to distinguish Muscovy from Russia. Nevertheless the two have been very closely interlinked, both geographically and historically. Expansion from the Muscovite core has in many ways taken on the appearance of being more of a continuous process than an abrupt transformation of one state into another. A number of geopolitical factors have influenced the character and direction of this massive expansion; while it took place right across northern Eurasia, the present examination will confine itself mainly to its implications for the western ecumene.

The absence of natural barriers is seen by Huttenbach as 'the basic condition of the Russian setting'.[5] The attainment of secure frontiers has always had a high priority in Russian foreign policy and their very elusiveness injected greater urgency into the search for them. Even Berdyaev, while seeking metaphysical answers as to the meaning of Russian history, found it difficult to ignore the overwhelming significance of geography. It was the urge for self-preservation, he observed, which had forced the Russians to push off invaders and to entrench themselves firmly in their habitats.[6] The initial danger to the existence of both Kievan and Muscovite Russia came from the southern steppes, and security here was a matter of priority. For many centuries it proved impossible to do more than hold a tenuous line across the north of the okraina where for a time a Cossack proto-state had been established. Following the removal of the threat from the Mongols, the principal threat to Russia switched to the west and was only eliminated with the defeat of the Lithuanians and the Teutonic and Livonian Knights. That historic 'Drang nach Westen' attributed to the Russians can thus be seen as a territorial response to the 'Drang nach Osten' of her neighbours. By the beginning of the nineteenth century control had been secured over the whole of the natural amphitheatre lying between the Arctic, Baltic, Black and Caspian Seas. The frontiers had been moved forward to the three isthmuses linking them, representing lines as close as possible in Russian conditions to 'les limites naturelles'. However, Russia then crossed these isthmian frontiers to incorporate Poland in the west and Transcaucasia in the south, for reasons which were political rather than strategic.

Commerce is another factor which helps to explain territorial expansion. The desire for greater wealth had motivated the princes of Muscovy in their endeavour to stimulate Moscow-orientated commercial activity by securing political control over the principal river routes. The Muscovite core region is located just east of the Valdai elevation from which the principal rivers of the Russian plain radiate out to the Baltic, Black and Caspian Seas. By the end of the sixteenth century Russia consisted physically of the basin of the Volga together with the Valdai elevation itself, and it thus controlled the principal routeways from the Baltic to the Caspian (Figure 5.1). The great commercial cities of Novgorod, Nizhni Novgorod, Kazan and Astrakhan were located along this vital routeway. Moscow lies just to the south of it, but closer to the sources of the southwards-flowing rivers, in particular the Don and the Dniepr. It was thus well located to protect and control the commerce of the Russian plain. With the subsequent acquisition of the Volkhov-Dniepr routeway, the central axis of Kievan Russ, Moscow secured control over the communication system of the whole of the Russian quadrilateral.

The seas into which the great Russian rivers flowed, and the coastlands around those seas, presented a more persistent problem of control. As the absence of natural barriers has been highlighted as the 'basic condition' of Russia, so the 'urge to the sea', springing from the desire to break out of its continental isolation, has frequently been cited as the fundamental theme in Russian history.[7] More specifically, the desire for warm water ports, ice-free and usable throughout the year has been identified as being the Leitmotiv in the process of expansion.[8] This drive has in fact had far wider economic, political and strategic motivation than the simple desire to possess a seaport might imply. It has centred on successive attempts to develop and modernise the country through exposure to maritime, and especially western, influences. Probing took place in all directions, but success was limited as a result of the tight political swathe wrapped around the Russian heartland from the Arctic Ocean to the Black Sea. This was held by the empires of Sweden, Poland-Lithuania and the Ottomans, all highly apprehensive of Russian moves into their respective spheres. The breakthrough finally took place in the Baltic, and this led to a fundamental geopolitical reorientation of the Russian state. Although ultimately so significant, this breakthrough was nevertheless at the time only a substitute for the more enduring territorial objective of expansion southwards. Although there were undeniably important commercial and strategic reasons for it, this Slavonic 'Drang nach dem Süden' was firmly grounded in Russian history.

Muscovite Russia claimed its legitimacy as successor state to both Kievan Russ and the Byzantine Empire. The idea of an indivisible whole with Muscovy as the state and Kiev as the 'soul' underlay both the unification of the Russkii and the 'gathering' of 'all the Russias' - the Rossiiskii - under the leadership of Moscow.[9] Together with the otchina - the patrimony theory - this concept constituted the principal justification for expansion. Later it came to be applied to the Slavonic peoples as a whole and was gradually realised with the defeat of Lithuania, the partition and absorption of Poland and the Slavophile and the Pan-Slav movements of the nineteenth century. It was sustained by a powerful crusading spirit engendered by the long wars against the Mongols and subsequently directed against the Catholic Poles and the Muslim Turks. Supported and encouraged by the Orthodox Church, the Tsars clung to the belief that the expansion of their state represented the fulfilment of the Divine Will. 'They conceived their role in religious terms', said Huttenbach and

their expansionism was the result of 'the zeal born of religious messianism'.[10] The Church's role being essentially
one of support for, rather than rival to, the secular power,
the situation became one of 'a complete caesaro-papism where
Church and State became synonymous'. In this world-view of
the state the Russian people, possessing greater wisdom and
piety than others, were the custodians of the true faith and
the agents for the reunification of Christendom. The Fall of
Constantinople had left Russia as the last champion of Orthodoxy, a beleaguered stronghold surrounded by heretics and
heathens. Out of this perilous situation came the sixteenth
century eschatological myth of Moscow as the 'Third Rome',
the new Eternal City which was the true heir to Rome and
Constantinople and successor to all Christian empires.[11]
The Rossiiskaia Imperiia was first holy and only second
Russian, and it was the religious impulse which lay at its
heart. It was an often reiterated objective of Russian policy
to repossess the Byzantine lands in the name of Orthodoxy.
'The Russians', said Martin Wight, 'regarded the Ottoman
Empire in Europe as being Christian terra irredenta'.[12]
While Orthodoxy thus provided the justification, the old core
region of Byzantium was also certainly coveted by the rulers
of Russia for many practical reasons of state, including
access to the sea, commercial advantage and the security of
their southern sea routes. The securing of control over
Constantinople and the Straits, the 'gateway to southern
Russia', was considered by Bowman to have been 'for
centuries ... one of the settled aims of intelligent Russians,
as well as of Russian rulers'.[13] However, all the diverse
reasons were grouped together under the banner of Holy
Russia, the successor state to Byzantium and the protector of
the Orthodox faith throughout the world. Undoubtedly this
religious impulse was a powerful one in its own right, but, as
with the justifying ideologies of the other dominant states
which have been examined, it was inextricably bound up with
the soil in which it had been nurtured and with the satis-
faction of the material needs which arose from the physical
conditions of its existence.

 The acquisition of the Ukraine, the Black Sea coasts and
the Caucasus were all accomplished as part of this process,
but no Russian penetration took place west of the Carpathians
or south of the Danube. Historically this area had attracted
Russian attention, particularly on account of the presence
there of both South Slavs and Orthodox Christians. Support
was given to Serb and Bulgarian independence movements
against the Ottomans, and the assistance to their embattled
South Slav kin gave rise to considerable popular enthusiasm
in Russia. To the Pan-Slavs Russia was the natural leader
and organiser of a great union of the Slav peoples which was
destined to contribute a 'new principle' to history. This union
was to include the historic lands of Byzantium and the more

ambitious and visionary among the Pan-Slavs saw it as being centred on Constantinople, 'Tsargrad', itself.[14] This city, it was argued, was best placed geographically to accomplish the great mission of the final triumph of Slav civilisation under the aegis of Russia. Not only would the Slavs then fulfil their ordained role as champions of Christianity against Islam, but they would also constitute a spiritual alternative to the 'Romano-Germanic' Europe of the west. In the Pan-Slav concept of the nineteenth century the messianic ideas of Holy Russia were fused with the Slavic Volksgeist to produce a universal mission. The Russian interest in south-east Europe was vigorously resisted by Austria and by the Ottoman Empire, the hegemonial states of Danubia and the Balkans respectively; in the later nineteenth century these two powers were supported by Germany and Great Britain, both increasingly fearful of a precipitate collapse of the flagging Ottoman Empire and the probable advance of Russia to Constantinople in fulfilment of her historic dream.

The southwards expansion to the Black Sea and the Mediterranean was thus fraught with difficulty and the major breakout from the land-locked core was westward to the Baltic where resistance was weaker. The opening of this 'window on the west', together with the transfer of the seat of political power to the estuary of the Neva, had two profound consequences. In the first place it opened Russia to contacts with northern rather than with southern Europe and this just at the time when the centre of power had definitively moved north of the Alps. What the historian Solov'ev called 'the wholesome and lifegiving influence of the ocean' was felt more strongly there than in the south where the Ottoman Empire was increasingly moribund and the Mediterranean Sea was becoming a backwater.[15] Secondly the transfer of the capital away from Moscow enabled the foundations of the new imperial system to be laid in territory far removed from the retarding influences of the old Russian core. The new capital was built on land only recently wrested from the Swedes and was to remain far less Russian in character than Moscow. 'A cross between Byzantium and Stockholm', said Nevill Forbes, 'it never was anything but foreign, as its founder wished it to be'.[16] From the outset it was planned as a neo-classical city, a large part of its people non-Russian and its aristocracy favouring the French over the Russian language. The autocracy itself came to typify this cosmopolitanism; Peter himself was known as the 'Germanised Russian' and his successor Catherine II as the 'Russianised German'. In the anarchist Bakunin's view, their one common characteristic was that they were all tyrants of foreign origin. 'At times', observed Kolarz, 'Russia was as little Russian as Austria was German'.[17] For the two hundred years during which St. Petersburg was the capital, the concept of Rossiiskaia Imperiia triumphed over that of Russkaia. This was demonstrated by the fact that little attempt was made to 'Russify' the Baltic acquisitions with their German-European culture, cosmopolitan cities and thriving maritime commerce. This was deliberate state policy since an important role had been assigned to them in the transformation of Russia from a backward landlocked state into a modern commercial and industrial one after the western model. One of the most important political imports from the west was what Kristof called 'the contagious example' of maritime imperialism, founded on commerce and justified as the vehicle for bringing progress and enlightenment to the dark parts of the globe. [18] This was to provide added justification for subsequent Russian expansion, mainly in central Asia and the Far East. However, on this the 'westerners' and the Pan-Slavs were opposed to one another. Like the two-faced Janus, as Herzen describes them, they were looking in opposite directions.[19] The former wished Russia to be one of the great powers of Europe hobnobbing on equal terms with the others, while the latter clung to the vision of her destiny as leader of a great Slav empire, holy and uncorrupted, in opposition to the 'Romano-Germanic' west. These two opposing ideas found physical expression in St. Petersburg and Moscow respectively. 'The two capitals', said Kristof, 'were symbols both of different ways of life and of different views as to what all of Russia should be like'.[20]

In the Russian Empire of the late nineteenth century with its vast territories, its multitude of nationalities and its variety of economic and social conditions, both Russian nationalism and Pan-Slavism were on their own inadequate to act as an unifying force. They could not provide a world-view sufficient to sustain so vast and heterogeneous a political structure. Even the idea of imperialism as a 'civilising mission' in the western sense made only a limited impact. What did gather them was the autocracy itself. The Tsar, father of his peoples and protector of Orthodoxy, was the ultimate symbol of the Imperial purpose in the world. In this way a paradoxical symbiosis of Muscovy, Third Rome and gatherer of the Russian lands, with St. Petersburg, focus of modernisation and 'window on the west' produced the cement which held this massive empire together.

This brings us to the link between the geographical and political character of the Russian state and it predilection towards expansion. 'In retrospect there seems something inevitable about the expansion', said Hatton, considering Muscovy's location in relation to the river communications of European Russia.[21] This factor, coupled with that absence of natural barriers which Hunczak thought of as being 'the basic condition of the Russian setting', endowed the expansion with what may be thought of as being a 'natural' character. The political core in the Mezhdurechie lies close to the hub of the radial river communications across European

Russia. The Valdai elevation, as Kristof has observed, was the cradle of Russian state power from which the Muscovite princes had been able to expand their territories with relative ease down the great rivers.[22] Thus Muscovy 'led by the rivers' extended ever outwards, finally seeking to attain control over the seas into which those rivers flowed.[23] Another factor was that the considerable physical homogeneity of the plain favoured Russian colonisation in order to alleviate the population pressure which, ever since the Mongol conquest, had built up in the deciduous woodland region. Especially favourable to Russian colonisation was the wooded steppeland stretching latitudinally from the upper Volga to the Urals and beyond, an area in which climate, soil and vegetation conditions bore close similarities to those of Russia itself.[24]

As to the political character of the state, the connection between absolutism and imperialism has been observed to be close, and this applies equally to the behaviour of the government and of those governed. Hunczak points out that colonisation preceded all other manifestations of Russian expansionism.[25] Facilitated and encouraged by a favourable coincidence of geographical conditions, colonisation was made more attractive still to the peasantry by the harshness of the Muscovite political system, especially after serfdom became official policy. The desire to escape to free lands out of the immediate reach of the Tsar and the boyars supplied a powerful motive for the colonisation of Siberia. In this way 'the Russian peasant preceded the state' and so helped accelerate official Russian expansionism.

In turn the official expansion itself was subject to few of the restraints which would normally be brought to bear on governments in more pluralistic and diverse societies. This gave the rulers of Muscovite Russia that 'political and tactical superiority that made territorial acquisition possible'.[26] As a method of increasing his power, Ivan IV created the oprichnina, a new class of loyal servants of the state which gave him a power base independent of the aristocratic boyars. From that time on the Tsars gathered around themselves a military-aristocratic class which identified closely with the autocracy as the principal source of its privileges and which regarded expansionist war as being a way of maintaining and strengthening its position. In these circumstances the pre-dilection of successive Tsars for warmaking leading to the acquisition of new territory suggests that they saw this as part of their role and relished the glory which accompanied it. Making war was the most heroic aspect of governing the state and provided a spectacular alternative to rational policy-making. Peter I lost perspective on strategic necessity and prolonged the war with Sweden in a vain and obsessive attempt to become master of the Baltic and to convert it from a Swedish into a Russian mare internam. Likewise Catherine

II, the 'Russianised German', relished her wars with the Ottomans, triumphantly crusading in a secular age to reinforce her precarious hold on her adopted country. The incorporation of the Kingdom of Poland after the Napoleonic Wars could hardly have been justified on practical grounds, but represented the final Russian triumph in the conflict with the rival Slav empire, begun two centuries earlier with the wars against the Lithuanians. Subsequent nineteenth century expansion was also in large part dictated by reasons of imperial grandeur and great power rivalry. Stoianovich points out that Schumpeter went so far as to say that Russian expansionism was 'a question of whim and fashion' undertaken entirely for the purpose of achieving 'sovereign splendour'. [27] This may indeed be a reasonable assessment of what went on in the mind of the autocrat, but Stoianovich sees it as being quite inadequate as a total explanation. It also fails to address the special features of Russian imperialism and the relation between geographical conditions and the direction which the expansion took.

Just Juel, a Danish envoy at the court of Peter I, clearly identified the connection between Russia's domestic backwardness and her imperial drive.[28] In this view expansion was an indication of the failure of successive governments to come to grips with the real problems faced by the country. 'Internal vacillation', said Schmidt-Hauer, was always followed by 'yet another expansionist drive'. A recurring theme in Russian history was that 'the Tsars, whether they acted as barbarian despots or as humane reformers, always increased Russian territory'.[29] Thus expansion, while certainly reflecting the appetite of the autocracy for Schumpeter's 'sovereign splendour' as an end in itself, also provided a distraction from the very real political, economic and social problems which arose out of the condition of the state. Military triumph was, in these circumstances, the easy option, and it was something which the autocracy, ever since the days of the early Tsars, had proved itself well equipped to achieve. Thus while providing a diversion from domestic problems it also diverted the energies of the autocracy into the pursuit of that elusive fata Morgana which appeared to beckon Russia towards a brighter future over the horizon.

'What is needed,' said Pleve in 1903, 'is a little victorious war to stop the revolutionary tide'.[30] Instead within eighteen months there came humiliating defeat at the hands of Japan, a relative newcomer to the ranks of the powers. The true extent of the weakness of the massive Russian Empire was then exposed to the watching eyes of the world. The Far Eastern diversion proved to be a final throw; it rebounded in the face of the autocracy and shook it to its foundations. The modernisation policy, hastily embarked upon in a desperate attempt to stem the rising tide of discontent, proved too little and too late to prevent the final collapse.

As has been seen, despite the massive scale of its
expansion, and the achievement of a hegemonial position over
the eastern part of Europe, the Russian Empire did not
become one of the dominant powers in the western ecumene,
although the whole thrust of the state was towards the attainment of such a role. The Russian Empire will now be examined against the model of the dominant state in an attempt
to explain this discrepancy.

The features of the Muscovite core region fit in well with
the model. Muscovy was a frontier state at the junction of the
free Russian lands, which paid tribute to the Mongols, with
those lands directly under the Mongol yoke. It was located
deep in the interior, remote from the centre of the parent
state and initially poor and backward compared to it. Its
structure was absolutist with pronounced theocratic elements
and it had an inherent propensity for aggression. Originally a
part of the parent state, it became independent first as a
result of the collapse of the parent state, and then of the
parent culture itself. The core state expanded towards a
hydrographic focus of routeways, and from there it was able
to exert control over the whole of the radial river system.
This was instrumental in generating the commercial wealth
which both increased the power at the disposal of the state
and underpinned its further expansion. This expansion took
place across the historic frontier and into the territory of the
opposing culture and, with the collapse of the latter, Russian
control was imposed over its western territories. Further
expansion in the same direction took the state away from its
heartland in the mixed forest belt and into the new environment of the Steppe, which in turn proved to be a source of
considerable agricultural and industrial wealth. The southerly
expansion was accompanied by colonisation by Russians and
brought the territory controlled by Muscovite Russia closer to
the macrocore of its parent culture. The continued expansion
of the Russian state was associated with pressures of external
origin challenging the pre-eminence of Muscovite Russia in the
Slav lands and even going so far as to threaten the existence
of the state itself. The expansionist thrust was then directed
into the lands of the ethnic-linguistic community of which
Muscovite Russia formed a part and over which it now asserted is position of dominance. By successfully uniting
almost all these eastern Slavs, Muscovite Russia produced the
circumstances in which a core nation could emerge out of the
ethnic fluidity which had preceded it. The Muscovites then
moved to absorb the lands of Kievan Russ, thus becoming its
successor state and inheritor of the role of protector of the
culture. The thrust of this expansion was along the rivers
into uninterrupted contiguous lowlands; this constituted the
geopolitical 'logic of unity'. This phase culminated in the
attempt to acquire a coastline and so, at last, to break out
from its 'claustrophobic isolation'.[31] The Arctic Sea having

proved to be of very limited value, the major breakout was to the Baltic, and a maritime 'window' was established there. At the same time the centre of political power was also trans-ferred to the Neva. Meanwhile continuing expansion to the south resulted in the creation of a second maritime 'window!, this time on the Black Sea. The 'natural' frontiers of the state thus became its coasts, together with the shortest isthmian lines joining them.

If those characteristics which pertain to the achieved position of dominance, which ipso facto do not apply to Russia, are excluded, then the Russian Empire has a 90 per cent concurrence with those characteristics itemised in Table 1. This is the same degree of concurrence as is found in Germany and higher than that of any of the other dominant states under consideration. Of those high-incidence character-istics which were used in the construction of the model, the Russian Empire also has 90 per cent, which is somewhat lower than Germany, the Ottoman Empire and Spain, but higher than either Austria or France. Looked at from the geopolitical viewpoint, Russia is thus very much a latent dominant state which nevertheless did not succeed in moving to Stage 3 of the model. In evaluating the geopolitical factors in this failure, it is necessary to examine the particular discrepancies which exist between Russian behaviour and the model.

Following her close initial conformity with the model, there then followed some significant divergences from it. It followed the model in defeating Culture B - the Mongols - and then reorientating back towards its parent state and absorbing it into its own proto-empire. It then followed this by expanding towards the macrocore of its parent culture. However, by that time, this macrocore had been absorbed into the dominant state of the new opposing Culture B - the Ottomans - and the macrocore of this Culture B had been transferred to that of the parent Culture A - the Byzantine Empire. This Culture B proved too powerful to dislodge, and Russia's breakout from continental isolation took place towards the emerging macrocore of the new sister-culture of Culture A centring on north-west Europe. While the reorientation towards this version of Culture A did a great deal to modern-ise the country, from the outset it relegated her to a rela-tively weak and subsidiary position. This was largely because, unlike the other reorientations towards a parent culture which have been examined, the macrocore of north-west Europe was in course of growth rather than of decline. As a result it proved far too vigorous to accept domination by any external power, and consequently Russia was able to play only a relatively modest and subsidiary role on its periphery. Both France and Germany were to make successive bids for hegemony in this new transalpine west, but the most that Russia could achieve was limited economic and political access to it. This represents a considerable deviation from Stage 2 of the model, and is geopolitically the point at which Russian aspirations to dominance were frustrated. However, the new Russian maritime core had many of the characteristics of an alternative to the north-west European macrocore of Culture A. Evidences of this were the powerful alternative world-view, the position of dominance east of the Baltic-Black Sea isthmus and the important role played by Russia as a peripheral counterweight to the hegemonial bids from the centre.

The dual orientation implicit in this situation then became the principal characteristic of Russian behaviour in the western ecumene. The Baltic core region became the political, maritime and commercial centre of the country. The south, centring on Kievan Russ developed into the country's major indigenous economic base; extending from west to east across southern Russia from the Dniestr to the Don and lying between the upper Volga basin in the north and the Black Sea-Caucasus divide in the south, it proved to be an area rich in both organic and inorganic resources. Between these two regions lay the old Muscovite core, now forsaken, but still retaining its hold as the major centre of the old Russian religious and cultural heritage while St. Petersburg remained in many ways alien. As Bater put it, 'St Petersburg served as Russia's administrative heart, Moscow its soul'.[32]

Thus while Russia conforms to a very high degree with the geopolitical model of the dominant state, it does deviate from it in a number of fundamental ways, and this discordance underlies the Russian Empire's failure ever to become more than just one of the great powers, influential, but never dominant. As has been observed the cultural macrocore towards which Russia moved after her success in becoming the inheritor state to the Mongols and to Kievan Russ, was held by the Ottoman Empire, an alternative universal state which had become far too powerful to dislodge. This was Russia's fundamental failure because control over the cultural macrocore would have brought with it control over major routeways, coastlands and maritime commerce, as well as a secure strategic position in the south. The nineteenth century Russian idealists saw Constantinople as being the natural centre of an Empire which would inevitably attain a position of dominance in the western ecumene. Instead Russia had to make do with access to a substitute cultural area in which she was regarded as being a half-oriental intruder viewed with suspicion by those powers located closer to its centre. When the alien power which had achieved domination over her own cultural macrocore became too weak to hold it any longer and Russia at last had a chance to make good her earlier failures, she was then frustrated by the combined energies of the European powers. They were by then fearful of the implications of the realisation of Russia's historic dream for the security of their own global imperial structures. The dual orientation thus perpetuated, far from diminishing Russia's expansionist drive appears to have accelerated it, but at the same time to have made it more arbitrary and capricious. To observers such as Schumpeter it appeared to be motivated more by that 'whim and fashion' leading to 'sovereign splendour' than by any more coherent purpose. This became most in evidence in Russian behaviour outside the western ecumene. The dual orientation was multiplied into a triple or quadruple one with the expansion southwards towards India and eastwards at the expense of the disintegrating Chinese Empire. The consequent dissipation of the country's energies made success in any one area all the more difficult to attain.

Certain of its internal geopolitical characteristics also detracted from the strength of the Empire and its ability to achieve a position of dominance. One of these was the incompleteness of the 'gathering' under Muscovite Russia with the implications of this for the emergence of a strong core nation. While by the nineteenth century dominion over the east Slavs had been accomplished, there remained much restiveness among them. Catholic Poland and Lithuania had been alternative foci around which the unity of the Slavs could have been achieved, and the triumph of Orthodox Muscovy was by no means universally acclaimed. The Slav sister nations, in particular the Ukrainians, White Russians, Lithuanians and Poles kept alive their own sense of cultural and historical identity and of the alternative possibilities which existed for them. Thus, even in its relationship with the other Slav peoples, the Muscovite enterprise always had a strong imperial quality about it. That total unity towards which Muscovy aspired remained incomplete, and the core nation did not stand out with such clarity as a cultural phenomenon as did Castile, France or even Kleindeutschland. The transitional quality of its boundaries with the other Slav peoples served to weaken its own identity and thus to impede its capacity for acting independently. Pan-Slavism was a later attempt to attain this elusive unity, but its exclusivity made it an unsatisfactory tool for the rulers of a large multi-ethnic empire. The only things which had a chance of holding the Empire together were Orthodoxy and autocracy, God and the Tsar, but in the end they too failed.

Another characteristic militating against success was the nature of the capital city itself. To Forbes, plainly no lover of St. Petersburg, it was 'cold, sunless, tragic, mysterious, dank and gloomy like the forests which surround it', and he concluded that it had exercised 'a sinister and unwholesome influence on Russian history'.[33] The large army of imperial functionaries which was obliged to live in the capital may well have agreed with Forbes' strictures on the climate, but in explaining its effects on Russian history it was its location which was a far more salient factor. Since the political centre was both geographically so peripheral and culturally so alien from the Empire as a whole, the problems of imposing central-isation and homogeneity on such far-flung and heterogeneous dominions were exacerbated. This weakness at the heart of the state was made still worse by the country's economic underdevelopment at a time when the possession of a strong industrial base was becoming essential to the wielding of international influence. The southern economic core region consisted largely of the extension eastwards of the carbon-iferous belt which stretches across the northern fringes of Hercynian Europe from northern France to the Carpathian mountains, but the Russian section of this zone lagged far behind the developments which had taken place further west, particularly in Germany. The principal ingredient lacking in Russia, but present further west, was the existence of a macrocore as a focus of cultural accumulation and technical and scientific innovation. In addition to this, in southern Russia the main centres of industry were remote and dis-persed and those contacts which were the essence of progress in western Europe were thus inhibited.

Whether the Byzantine macrocore, if acquired, could ever have served as such a focus is bound to be speculative. The occupation of the macrocore of the parent culture by the power of the opposing culture is an unique phenomenon in this context. However, the Marmara region has many geopolitical characteristics in common with the other macrocores of the western ecumene. Had the Ottoman presence been removed, it is quite reasonable to suppose that the pre-Ottoman world would have been revived and the Eastern Mediterranean-Aegean-Black Sea linkages resumed. This would have made the Marmara region once more the geographical focus of communication and contact between southern Russia on the one hand and the Balkans, Greece and Anatolia on the other. In such a scenario, the Pan-Slav vision of 'Tsargrad' as the centre of a Russian Empire which would have attained dominance in the western ecumene seems not too far-fetched.

The reality of the Russian situation was pointed out by Turgenev who evoked the image of 'Holy Russia' sunk into a drunken stupor, 'her forehead at the pole, her feet in the Caucasus'. Indeed the facts of geography appeared to con-spire with world events to keep the Russian Empire well below both her political and her economic potential. When industrial development did come it was as a result of the injection of western capital and technology. It thus served to weaken further the real power, and ultimately the political autonomy, of the precarious tsarist regime.

To sum up, the geopolitical characteristics of the Russian Empire which do not conform to the pattern of the dominant state give an indication of why Russia failed to achieve that status. Otherwise, as has been seen, the in-cidence of dominant state characteristics is actually higher in Russia than in any of the states examined except for

Germany. Although Russia only attained a limited regional hegemony, she is actually geopolitically more similar to the dominant states than to those other states in the western ecumene which attained a limited regional hegemony. The three most important such states are Sweden, Poland and Italy which attained hegemony over the Baltic, eastern Europe and the eastern Mediterranean respectively. However, each of them went on to claim for itself a more comprehensive role, Sweden claiming to be the hegemonial power of northern Europe, Poland the natural leader of the Slavs and Fascist Italy grandly asserting its position as successor state to the Roman Empire in the Mediterranean. However, not one of them was successful in the fulfilment of the wider destiny which it envisaged. Underlying this is the fact that the overall incidence of dominant state characteristics in these three is much lower than in those which actually did attain a wider position of dominance. This is so even allowing for those particular characteristics which relate to the actual position of dominance which naturally do not apply to them. Particularly significant discrepancies relate to the position and behaviour of the original and historic core regions, the location of the centre of power, relationship to other culture areas, the assertion of control over the parent macrocore and the colonisation of newly acquired territories. Despite the enormous size of Russia's territory compared to these lesser hegemonial states, the extent of the Russian achievement in the western ecumene is more akin to theirs than to that of the five dominant states. However, as has been demonstrated, her geopolitical characteristics on the other hand are closer to those of the dominant states than to those of the states which achieved only limited regional hegemony. Russia's physical and human resources coupled with her ambitious world-view single her out as being a power likely to achieve a dominant position, and the reasons for her failure to do this can be traced to anomalous characteristics in her geopolitical structure.

In his examination of the origins of Russian imperialism, Huttenbach concluded that the country's expansion did not possess such distinctive historical qualities as to 'set it apart from the expansionist forces operating elsewhere in Europe'.[34] Yet the Russian Empire was, and remained, the object of considerable apprehension to the peoples of the western world. Despite the westernisation of its aristocracy and the efforts of the autocracy to establish and maintain its position as a European power, its image remained alien and half-Asiatic, partly because of the geographical fact of distance. Indeed, its absolutist government with theocratic undertones had made it into a kind of anti-symbol to nineteenth century liberals in the west. There was also a very real fear in the chancelleries of Europe of a country possessed of such power, which it had already used to play a

Figure 5.2: Geopolitical Characteristics of the Russian Empire West of the Urals in the early Twentieth Century

decisive part in the affairs of the continent. The fears were also grounded in western perceptions of the geopolitical character of Russia. First there was its sheer size, far larger than all the other powers of Europe combined, a fact conveyed only too clearly by the new and colourful political atlases. 'Upon a glance at the map', wrote the American admiral A.T. Mahan in 1900, 'one enormous fact immediately protrudes itself upon the attention - the vast, uninterrupted mass of the Russian Empire, stretching without a break in territorial consecutiveness from the meridian of western Asia Minor, until to the eastwards it overpasses that of Japan. In this huge distance no political obstacles intervene to impede the concentrated action of the disposable strength'.[35] What Mahan observed at the end of the nineteenth century was the result of that propensity for expansion over immense areas of contiguous territory which had since the seventeenth century made the Russian Empire the largest territorial state in the world. 'The truly enormous scale' of practically uninterrupted expansion had made Russia in Halecki's view 'an absolutely unique case'.[36] Furthermore there was the sheer inaccessibility of the continental heart of the country. It was possible for the Russians to retreat endlessly into its vastnesses, drawing the enemy on to exhaustion in the harsh and unfamiliar immensities. Finally, there was the vast resource potential which already in 1900 looked greater than that of any other part of the world.

This widespread fear of Russia and of Russian intentions was the seedbed of the geopolitical ideas of Halford Mackinder. His 'Pivot' theory was first propounded in 1904[37] and was later modified and expanded into the 'Heartland' theory.[38] The Pivot was the name he gave to the centre of the Eurasian landmass stretching from the Himalayas to the Arctic Ocean. Its outline was delineated by the watershed between the rivers draining to the ocean and those which drained into the interior of Asia or into the ice-bound northern seas. This, according to Mackinder, was an area completely out of the reach of maritime power and was consequently, in the conditions of the time, invulnerable. The Pivot included the area which drained into the Caspian Sea and therefore the whole of the Volga basin. However it did not include those lands which were drained by rivers flowing into the Baltic and Black Seas which, unlike the Caspian, have access to the oceans. The western boundary of the Pivot thus lay across European Russia, and this suggested evidence of an inherent maritime-continental duality in the Russian condition (Figure 5.2). The old Muscovite core was located within the Pivot, while the maritime core on the Neva lay outside it in the Marginal Crescent. Mackinder's theory thus put Russia into a unique geopolitical position and ordained for her a special place in world history. He contended that the central theme of history was the conflict between land power and sea power, and over the millennia the balance of advantage had oscillated between the two. The ultimate conflict, towards which history had been moving, would be between that land power which possessed the Pivot and the maritime power which was able to dominate the Marginal Crescent. In what Mackinder referred to as the 'Columbian Age' the maritime powers had possessed the advantage of the relative ease and cheapness of communication by water as compared to the cost and difficulty of communication on land. The coming of trans-continental railways however was in course of transforming the situation, making possible the opening up of the land masses. In these circumstances the power which controlled the Pivot would have access to immense natural resources and these, along with its strategic location and its invulnerability, gave it a conclusive advantage in the achievement of world dominance.

This theory was the basis for much subsequent geopolitical thinking about the position of Russia as a great power. However, the relationship between it and the geopolitical characteristics examined in this chapter is a more subtle one. Certainly there existed the vast · and relatively homogeneous territory and the propensity of Russia to extend her control over it. The Russian historical core area was strategically located within the Pivot, and the inability to subdue the Russian core was the most important military factor in the failure of the French bid to achieve lasting dominance in Europe. Russian conflicts such as that with the Polish-Lithuanian Commonwealth, the Northern War with Sweden and the Crimean War with Britain and France were largely confined to the peripheries of the Empire and did not violate the core. Certainly too the country's physical resources, in particular those known to exist in the Ukraine-Caucasus belt, were on an immense scale and at the beginning of the twentieth century had scarcely been tapped.

However, despite the presence of what Mackinder saw as being special explanatory factors, the country's wider geopolitical characteristics and patterns of behaviour are as has been observed not unique, as Mackinder's theory would maintain, but very similar to those of the other great powers of the western ecumene. The one feature of Russia which set her apart from the others was that, despite manifest aspirations in that direction, she did not achieve the dominance which they did. The reasons for this have been traced to the geopolitical conditions of her existence; these conditions therefore must have been in certain ways deleterious to Russian aspirations, a conclusion which is the opposite of what Mackinder himself asserted.

Central to Mackinder's argument was the contention that Russia possessed unmatched invulnerability and physical resources, and we have already observed elements of truth in this assertion. But, though Mackinder may have been correct in claiming that the Pivot was invulnerable to maritime power such as that exercised by Great Britain, the Russian historical core within the Pivot had been occupied briefly by the French, who were another land power and had not been deterred by the interior drainage system, a factor central to Mackinder's delineation of the Pivot. What ultimately defeated Napoleon was the unsuperable logistic problem of servicing an army of 600,000 men at such a distance from their supply bases and in such inclement climatic conditions. That inviolability which Mackinder had seen as being a particular feature of the Russian core was, in any case, far from being unique among the dominant states. The historic core regions of both Spain and the Ottoman Empire, the Meseta and the Anatolian plateau, had remained unconquered by adversaries during their periods of dominance. As to the question of physical resources, it is undeniable that Russia was well endowed and had ample potential to develop a strong industrial base. However, the known resources were located almost entirely outside the Pivot, and the extent of that wealth was still at this period largely speculative.

Thus the actual geopolitical evidence does not support the widely-held view that the Russian Empire was a state possessing unique power combined with abnormally aggressive intentions. The reality of the time was one of considerable weakness and, very soon after Mackinder delivered his 1904 paper, this was made very clear to the world by the Russo-Japanese war. The thesis does give backing to the idea that Russia had the potential to become a dominant power, possibly even the next power to achieve domination over the western ecumene. However, this does not alter the fact that the principal geopolitical characteristics which distinguish Russia from those powers which had achieved a dominant position were still operative at the beginning of the twentieth century. They included particularly the incompleteness of core nation formation, the failure to dominate the cultural macrocore and the movement towards a substitute, the underdevelopment of the industrial base and the inadequacies of the attempts to exercise political power from the peripheral core. These special features put Russia into a weak position as compared to Germany, despite the fact that otherwise the two countries possessed almost identical geopolitical characteristics.

A new element in Russian geopolitical behaviour became evident in the later nineteenth century, and this was the reorientation towards Siberia. Russia came more and more to behave as a Eurasian power endeavouring to develop her eastern possessions in order to redress the balance after her failures in the west. This was, of course, not so much an entirely new direction as a return to the direction of pre-Petrine Russia when the rulers of Muscovy had marched eastwards as successors to the Mongol Khans. While formerly it had brought Russia into contact with the Chinese, it now brought her up against the far-flung dominions of Britain on the other side of the world. It was above all in Britain that Russia at this period came to be viewed as the 'anti-symbol' and principal danger to the stability of the world order. While in Russia the reorientation represented a move of interest from the Marginal Crescent back into the Pivot, in nervous imperial Britain it was perceived as the expansion of the Pivot state outwards towards the Asiatic parts of the Marginal Crescent where Britain still retained her fragile pre-eminence. Whichever way one looked at it, it did not indicate a renewed bid by Russia for a dominant position in the western ecumene, but a turning away from it. If the Russian Empire were to be saved, before the gathering political stormclouds made it too late, then it was over the eastern horizon that this appeared to remain a possibility.

NOTES AND REFERENCES

1. R.J. Kerner, *The Urge to the Sea: The Course of Russian History* (University of California Press, Berkeley, 1942), p. 35.

2. H.R. Huttenbach, 'The Ukraine and Muscovite Expansion', in T. Hunczak (ed.), *Russian Imperialism from Ivan the Great to the Revolution* (Rutgers University Press, New Brunswick, New Jersey, 1974), p. 168.

3. R. Charques, *A Short History of Russia* (English Universities Press, London, 1959), p. 14.

4. W. Kolarz, *Myths and Realities in Eastern Europe* (Lindsay Drummond, London, 1946), pp. 48-9.

5. H.R. Huttenbach, 'The Origins of Russian Imperialism', in T. Hunczak (ed.), *Russian Imperialism*, p. 20.

6. N. Berdyaev, 'O Vlasti Prostranstv nad Russkoi Dushoi', in *Sudba Rossii, Opyty po Psikhologii Voiny I Natsionalnosti* (Moscow, 1918).

7. R.J. Kerner, *The Urge to the Sea*, p. 62.

8. N.J.G. Pounds, *Political Geography* (McGraw-Hill, New York, 1963), pp. 78-9.

9. W. Leitsch, 'Russo-Polish Confrontation', in T. Hunczak (ed.), *Russian Imperialism*, p. 163.

10. H.R. Huttenbach, 'The Origins of Russian Imperialism', p. 27.

11. This idea emanated from a sixteenth century monk, Philotheus, who proclaimed, 'All Christian empires are united within yours, for two Romes have fallen, but the third stands and there will be no fourth. As the great prophet (David) has foretold, your empire will never pass into the hands of others'. Quoted in L. Poliakov, *The Aryan Myth: A History of Racist and Nationalist Ideas in Europe* (Chatto and Windus and Heinemann, London, 1974), p. 106. The idea has had considerable influence on the interpretation of Russian im-

perialism. Sarkisyanz sees it as being a 'simplistic convention' tending to obscure other more real causes of the phenomenon. See E. Sarkisyanz, 'Russian Imperialism Reconsidered', in T. Hunczak (ed.), *Russian Imperialism*, p. 52.

12. M. Wight, *Power Politics*, 2nd ed. (Royal Institute of International Affairs and Penguin, Harmondsworth, 1986), p. 307.

13. I. Bowman, *The New World*, 4th ed., (Harrap, London, 1928), p. 410.

14. T. Hunczak, 'Pan-Slavism or Pan-Russianism', in T. Hunczak (ed.), *Russian Imperialism from Ivan the Great to the Revolution* (Rutgers University Press, New Brunswick, 1974), p. 92.

15. See the outline of the conflict of ideas between Danilevskii and Solov'ev on the whole question of Russia's true destiny in L.K.D. Kristof, 'The Russian Image of Russia', in C.A. Fisher (ed.), *Essays in Political Geography* (Methuen, London, 1968), p. 373, note 1.

16. Quoted in R. Beazley et al., *Russia from the Varangians to the Bolsheviks* (Clarendon Press, Oxford, 1918), p. 245. Peter was actually more influenced by the Dutch. The original name given to the new capital was 'Pieterbruk' and it was only later that this was changed to its German version, St. Petersburg. It retained this name until World War I when, briefly, it was Russianised to 'Petrograd' for patriotic reasons.

17. W. Kolarz, *Myths and Realities*, p. 43.

18. L.K.D. Kristof, 'The Russian Image of Russia', in C.A. Fisher (ed.), *Essays in Political Geography* (Methuen, London, 1968), p. 368.

19. Referred to in W. Kolarz, *Myths and Realities*, p. 107.

20. L.K.D. Kristof, 'The Russian Image of Russia', p. 366.

21. R.M. Hatton, 'Russia and the Baltic' in T. Hunczak, *Russian Imperialism*, pp. 106-7.

22. L.K.D. Kristof, 'The Russian Image of Russia', p. 361.

23. S. deR. Diettrich, 'Some Geographic Aspects of the Russian Expansion', *Education*, Feb. 1952, p. 6.

24. R.E.H.Mellor, *The Soviet Union and its Geographical Problems* (Macmillan, London, 1982), p. 33.

25. H.R. Huttenbach, 'The Origins of Russian Imperialism', p. 21.

26. W. Leitsch, 'Russo-Polish Confrontation', p. 128.

27. T. Stoianovich, 'Russian Domination in the Balkans', in T. Hunczak (ed.), *Russian Imperialism*, p. 201.

28. C. Schmidt-Hauer, *Gorbachev - The Path to Power* (Tauris, London, 1986), p. 23.

29. C. Schmidt-Hauer, ibid., p. 31.

30. Quoted in H. Seton-Watson, *The Decline of Imperial*

Russia (Methuen, London, 1964), p. 213.

31. W.H. Parker, *Mackinder. Geography as an Aid to Statecraft* (Clarendon Press, Oxford, 1982), p. 190.

32. J.H. Bater, *St. Petersburg* (Arnold, London, 1978).

33. Quoted in R. Beazley et al., *Russia from the Varangians to the Bolsheviks*, p. 245.

34. H.R. Huttenbach, 'The Origins of Russian Imperialism', p. 21.

35. A.T. Mahan, *The Problem of Asia* (Little, Brown & Co., Boston, 1900), p. 24.

36. O. Halecki, 'Imperialism in Slavic and East European History', *American Slavic and East European Review*, Vol 11, 1952, pp. 1-26.

37. H.J. Mackinder, 'The Geographical Pivot of History', *Geographical Journal*, XXIII, 1904, pp. 421-37.

38. H.J. Mackinder, *Democratic Ideals and Reality: A Study in the Politics of Reconstruction* (Constable, London, 1919).

SIX

THE SOVIET UNION: SOCIALIST COMMONWEALTH
OR NEW IMPERIAL STATE?

'Russia is a peculiar country,' said Valeriy Tarsis, 'birthday
party one day, funeral the next - that's the whole of our
history'. In 1917 after three catastrophic years of war it was
the funeral which came first. The Empire which had been
ruled for three hundred years by the Romanov dynasty
collapsed in chaos, and the unexpected birthday party of the
new state was soon to take place amongst its ruins. Following
a brief and increasingly chaotic interval of something
which resembled bourgeois-democratic government, Lenin's
Bolsheviks assumed control and the Soviets of Workers and
Peasants, which had made a brief appearance in 1905, became
the foundation of the new order. The fledgling Soviet Russian
state was immediately faced with a set of problems so basic as
to call its continued existence into question. Principal amongst
these were the demands for independence coming from the
former subject nationalities, widespread economic disruption,
civil disorder and mounting opposition from counter-
revolutionary elements together with the Interventions by the
western powers in support of these 'Whites' against what they
perceived as being subversive 'Reds'. The new Soviet state
thus faced powerful opposition both inside and outside its
frontiers and was soon forced to employ the most draconian
methods to protect its interests. It was not until five years
after the Revolution which had brought the Soviets to power
that the Union of Soviet Socialist Republics was finally proclaimed.

The international situation which had been inherited by
the Soviet regime in the fourth year of World War I could
hardly have been a more serious one. A large slice of the
west of the country, including Poland and Lithuania, was
under enemy occupation. The German response to the chaos
which they had done so much to precipitate was to mount yet
another offensive while at the same time proposing peace
negotiations. The Soviets were at first unwilling to be in any
way associated with the foreign policy of their predecessors,
and were certainly not prepared to join the western powers in

what Marxist doctrine proclaimed to be a capitalist-imperialist war. However, when a renewed German advance threatened Petrograd, Trotsky, the first Soviet Commissar for Foreign Affairs, was forced to come to the conference table. In March 1918 the Treaty of Brest-Litovsk dealt a savage blow to Russian territorial power. Poland, the Ukraine and the Baltic nations were resurrected as independent states and Russia was pushed back eastwards beyond her pre-Petrine frontiers.

Had this treaty been fully implemented it would have spelt the end of Russia as a great European power and imposed German hegemony over the North European Plain eastwards to the Black Sea and the Don. However, six months after the Treaty was signed Germany was finally defeated on the western front, and Russia was reprieved. The consequent territorial settlement in the east largely conformed with the policy of the western powers at Versailles and with the stated Wilsonian principle of national self-determination. While both France and Britain subscribed to this policy, their underlying aim was the creation of a cordon sanitaire from the Baltic to the Mediterranean designed to prevent a resumption of either the German 'Drang nach Osten' or the Russian 'Drang nach Westen'. In accordance with this a number of new states were established in eastern Europe and the existing ones were strengthened. A wide band of territory stretching from the Arctic to the Black Sea along the western flanks of the defunct Russian Empire was never allowed to become part of the new Soviet Union (Figure 6.1). Poland was resurrected and the composite west Slav state of Czechoslovakia was brought into being. Estonia, Latvia, Lithuania and Finland also became independent states. Finally, the province of Bessarabia was detached and given to a much enlarged Romania. While these tremendous changes were sanctioned and guaranteed by the western powers, in many cases the nations of the region took matters into their own hands in advance of official international agreement. In particular the Poles, once again free after a century and a half of subjugation, aspired to the establishment of a large and quasi-imperial Polish state, commensurate with what they considered to be their historic rights. The British government proposed that the eastern frontiers of the resurrected Polish state should follow a line separating those areas in which the Poles formed the dominant population group from those in which the Ukrainians and Byelorussians were in the majority. This 'Curzon Line'[1] was rejected by the Poles as being totally inadequate, and they went on to lay claim to large territories within the former Russian Empire by virtue of what Lloyd George called 'the conquering arm of their ancestors'.[2] Displaying their own particular brand of national messianism, they claimed a 'Polish irredenta' in the east and aspired to replace Russia as the hegemonial power of the region.[3] In pursuit of this ambition, they proceeded to invade Ukrainian territory and to push back the Russians. When a frontier settlement was finally reached between the Soviets and the Poles, it was considerably further east than the Curzon Line and proximated to the line of the Second Partition in 1793.[4] In addition Russia lost virtually the whole of her Baltic coastlands apart from a 'mere peep-hole on the Gulf of Finland'[5]; she lost also the middle Vistula basin, the western sections of the Dniepr and the Dniestr basins and the northern flanks of the Danube delta. As a result of all this, the Soviet Union was in a far weaker territorial position in Europe than the Russian Empire had been since the beginning of the seventeenth century.

This post-Versailles territorial settlement was to endure for barely twenty years and was effectively brought to an end by the Nazi-Soviet Pact of August 1939. In it Germany and the Soviet Union agreed that the time had come for the reassertion of their joint hegemony over eastern Europe. The precarious cordon sanitaire, designed by the western powers to keep Germany and Russia apart, was torn down, and by June 1940 the Soviet Union had occupied those eastern regions of Poland largely inhabited by Byelorussians and Ukrainians and annexed all three of the Baltic republics. However, like the earlier Treaty of Brest-Litovsk, this was destined to be short-lived. In June 1941, after eighteen months of strained friendship, the Germans turned on the Soviets. Operation Barbarossa unleashed the Wehrmacht on the Soviet Union with the most devastating consequences. Within a year the Germans were in occupation of an enormous area west of a front extending diagonally across the country south-eastwards from Leningrad to Stalingrad and from there thrusting deeply southwards towards the Caucasus. The economic heart of the country, together with the greater part of its coal, iron, steel and cereal production, had fallen under enemy control. For a second time within a single generation the position of the Soviet Union was a precarious one. However, the massive German advance finally ground to a halt in front of Moscow and Leningrad, and at Stalingrad on the Volga the Wehrmacht suffered its first major defeat in Europe. It was from this extended limb of the front that the great roll-back began which did not come to a halt until the Soviets had conquered Berlin (Figure 3.4).

The Soviet Union was now for the first time in its existence a victorious power and in a position to improve upon the tightly drawn frontiers imposed a quarter of a century earlier. The territory acquired under the Nazi-Soviet Pact was retained, together with three small but significant additions in Karelia, East Prussia and Ruthenia. In Karelia the Soviets retained the whole of the isthmus between the Gulf of Finland and Lake Ladoga, gained from the Finns after the Winter War of 1939-40. This both strengthened her strategic position in the Baltic area and decreased the vulnerability of

Leningrad, which was previously only forty kilometres from the frontier. East Prussia was taken from Germany and partitioned between Poland and the Soviet Union. By allocating to themselves the northern half, the Soviets gained a foothold on the southern coast of the Baltic. Finally, with the incorporation of Ruthenia, formerly the easternmost province of Czechoslovakia, the frontier was moved for the first time west of the Carpathians and into the fringes of the Danube basin. The acquisition of the south-eastern territories of inter-war Poland meant that the whole of the Dniestr basin, including the south of the old Austrian province of Galicia, also became a part of the Soviet Union. Save only for the continued independence of Poland and Finland, the Soviet Union thus resumed and even improved upon the former frontiers of the Russian Empire (Figure 6.2).

The Soviet Union was also in a strong position to ensure that other post-war territorial arrangements in the eastern part of Europe were to its advantage. Her zone of occupation in Germany had been previously agreed with her two principal western allies at the Yalta Conference, but in the other areas liberated by the Red Army she increasingly took matters into her own hands. The transfer to Poland of those German lands east of the Oder-Neisse line and the expulsion of the German population from them followed a bi-lateral agreement with the new Polish government installed in the wake of the country's liberation by the Red Army. Similarly the detachment of the eastern territories of Czechoslovakia and Romania was undertaken unilaterally by the Soviet Union, and then justified on ethnic grounds. Few other territorial changes were made in those areas which fell into the Soviet sphere, but the Soviet presence remained and by 1947 the frontier of the area of Soviet dominance was solidifying. The significance of this was highlighted by Churchill's use of the term 'Iron Curtain' in 1946. Although frequently attributed to Churchill himself, the phrase had been used previously by leaders of the Third Reich to describe the junction between the zones of Soviet and Anglo-American domination which they foresaw following the collapse of Germany and the eclipse of Mitteleuropa.[6] The area of Soviet control was thus extended from the Russian Empire's old Baltic-Black Sea frontier, now re-established de jure by the Soviet Union, to a new forward de facto frontier stretching along the Baltic-Adriatic isthmus. This gave the Soviet Union a huge new quadrilateral of territorial power stretching from the southern coasts of the Baltic to the Balkans, and consisting of almost the whole of the basins of the Danube and the Elbe. It was a larger area of domination than the Russian Empire had ever succeeded in achieving. However, it represented a very temporary state of affairs since the extent of Soviet control within it was nothing like as total as at first appeared to the western powers. In 1947 Yugoslavia was able to detach herself from the Soviet

Figure 6.1: Western Frontier Changes after World War I

Figure 6.2: Western Frontiers and Sphere of Influence of the Soviet Union after World War II

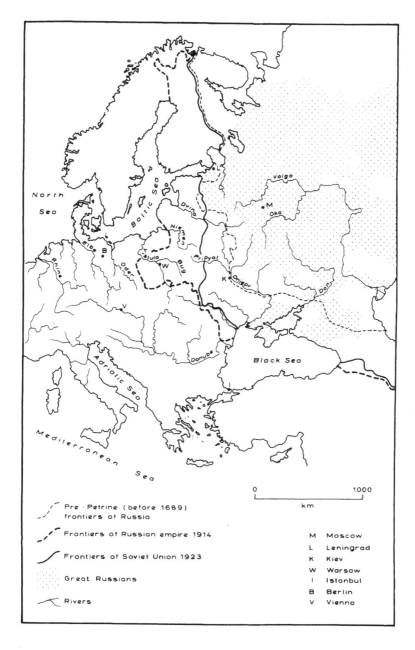

sphere, and from then she was to maintain a largely neutralist position. As a result, Soviet influence in the Adriatic region came to be confined to the mountainous and primitive outpost of Albania. In 1955 the Austrian Peace Treaty was signed and the Soviets withdrew from their zone in the east of that country. In the following year the Albanians refused to accept Khruschev's denunciation of Stalin, and proceeded to abrogate their own alliance with the Soviet Union, in consequence of which Soviet influence in the Adriatic and the western Balkans was virtually eliminated. These developments meant that the area of Soviet pre-eminence shrank back from the forward Baltic-Adriatic line to a new Baltic-Black Sea line running roughly parallel to the Soviet frontier and forming a clear buffer zone around it at a distance of between 500 and 700 kilometres (Figure 6.2). The six countries of which this inner cordon was composed were then bound ever more closely to the Soviet Union and came to be referred to contemptuously as 'satellites' to denote their essentially subordinate status.[7] Soviet power was exercised through overwhelming military preponderance, over-all political and ideological conformity and close economic interdependence. It operated within the three overlapping institutional structures of Cominform (1947), the Council for Mutual Economic Assistance - Comecon - (1949) and the Warsaw Pact (1955). While the headquarters of the two latter organisations were located in Moscow, Cominform was initially established in the Yugoslav capital, Belgrade, and its enforced displacement to Bucharest was a major blow to Moscow-led communist solidarity.

While the Soviet Union had an overwhelming pre-eminence throughout this cordon it was not translated into a position of absolute supremacy. The extent of Soviet control was from the outset limited by pre-existing conditions within the area, including its ethnic and cultural diversity, varying levels of economic development and divergent and often antagonistic political histories. These conditions eventually combined to deny the Soviets that close unity under their leadership which they clearly sought to attain.

The state which came into existence in 1923 was, from a Soviet perspective, sui generis - a new political form unlike anything which had existed before in history. It was supposedly founded upon the scientific theory of socio-economic development enunciated by Marx and developed by Lenin. Since it represented a new stage in the historical process its relationship to the discredited Russian Empire was seen as being purely a chronological one. The new state was designed to be nothing less than the forerunner of an entirely new type of society based initially on socialism and leading, through the iron laws of historical development, to communism, the ultimate form of human social organisation. This was all to take place according to a predetermined and largely immutable timetable. However, from the outset there was a problem of fitting theory to reality. According to Marx, the revolution which was destined to set in train the next stage in the world process would take place in the advanced capitalist west. Russia, still in Marxist terms in a pre-capitalist stage of development, had thus to pass through the capitalist stage before being able to attain socialism. As it was, the march of events did not wait for theory to become reality. In 1917, as a result of the precipitate collapse of the Russian Empire and the inability of the bourgeois-democratic government to address itself adequately to the immense problems which the country faced, Lenin's Bolsheviks were forced to intervene prematurely and so to bring about the Revolution sooner than the unfolding of the Marxist historical process would have dictated. The new state thus came into being in the most unpropitious circumstances from both the theoretical and the practical standpoints.

A far more serious deviation from the Marxist historical scenario was that the predicted revolutions did not then successfully take place in the advanced capitalist countries of the west, and the first socialist state found itself virtually isolated in a hostile capitalist world. This posed the immense practical problem of how the Soviet Union could continue to survive, as well as the ideological one of how it could continue to fulfil the proselytising role which was the main justification for its existence. Stalin's answer to this dilemma was the implementation of a policy of 'socialism in one country'. This entailed forcing the pace of agricultural and industrial development by means of a series of ambitious economic plans. The twin objectives of this sweeping pro-gramme were the laying of the foundations of communism by the modernisation of the country and the strengthening of the economic base so that the Soviet Union would be in a position to resist the predicted onslaught from the capitalist world. According to Stalinist thinking, victory over the enemies of socialism had to be attained and national security assured before the real building of communism could be resumed. The forced development of a largely backward and agrarian country did have the effect of converting it within a decade into a heavy industrial giant, and in World War II it was in a far stronger position than had been the Russian Empire twenty-five years earlier to resist and eventually to defeat the forces of Europe's most formidable military and economic power. In Soviet ideological terms the Great Patriotic War was the defeat of militaristic capitalism by the working classes in arms; thus the objectives of the Revolution were preserved and the historical process could be resumed.

While from the Soviet historical perspective the USSR is a totally unique and pioneering phenomenon, from the geopol-itical perspective this contention calls for closer examination. The geopolitical proposition which has been advanced in this book is that a state is founded not so much upon an abstract ideal as upon an objective physical reality. The occupation of a particular geographical space cannot therefore be considered as being merely a fortuitous and relatively unimportant factor. Rather it comes to be seen as a fundamental condition of the existence of the state, and one which moulds its nature, ideals and patterns of behaviour. Past state ideals have been born and refined in distinct sets of geopolitical conditions. In the light of this, despite Soviet contentions, it is reasonable to consider the proposition that the Soviet Union is, like the states already considered, less the pure embodiment of an ideal unique in history than a territorial state in whose particular spatial conditions a special world-view has emerged. Soviet assertions need to be tested against the nature of the territorial reality in which the Soviet Union has been created. This will be done with a view to ascertaining the extent to which its 'socialist reality' is also accompanied by a 'socialist geopolitical reality'. If the existence of such a phenomenon can indeed be established, then it is necessary to ascertain the extent to which this differs in kind from the other geopolitical realities which have so far been encountered. Since certain patterns of behaviour have been identified as being linked with particular geopolitical characteristics, it is reasonable to extrapolate from this that the geopolitical characteristics of the Soviet Union may be expected to throw light on Soviet behaviour. If it turns out that there is indeed an unique set of characteristics, then this will support the contention that the Soviet Union is, after all, as unique a state as it has always claimed to be. If, on the other hand, it proves to possess a set of characteristics common to other states, then this will point to the categorisation of the Soviet Union among such states, with the expectation of the same or similar patterns of behaviour.

Fundamental to the consideration of this is the question of the relationship of the characteristics of the Soviet Union to those of its predecessor, the Russian Empire. The former owners of the imperial estate, as it were, having been found wanting, the estate employees decided to remove them and then to declare themselves the inheritors of the property. However, revolutionary as this was, it did not in itself revolutionise the geography of the inheritance. It is important to consider, therefore, how revolutionary the subsequent transformation has been in the geopolitical sense.

Altogether there are four possible explanations of the nature of the geopolitical structure of the Soviet Union in relation to the preceding Empire. The first one is, so to speak, creationist, that is that the structure came into being at the moment that the Soviet Union was established; it would thus be an entirely new phenomenon, unrelated to that which had existed on the same territory previously. The second explanation is that the characteristics of the Russian Empire persisted alongside those of the new Soviet Union. If this proves to be the case, then there are, in geopolitical terms, two separate states occupying the same geographical space, one of them imperial and the other one post-imperial. The third possibility is that the characteristics particular to the Soviet Union have fused with those left over from the Russian Empire to produce a hybrid state deriving something from each of them. The final possibility is that the characteristics of the Soviet Union are fundamentally the same as those of the Empire it replaced. If this last possibility proves to be correct, then what we have, in geopolitical terms, is essentially an imperial state of the traditional type which is, in effect, masquerading as being something new and different. Whichever of these possibilities proves to come nearest to the truth, the one time existence of the geopolitical structure of the former Russian Empire within largely the same territory as that of the Soviet Union is an historical fact which is not open to question. There exists an historical political geography, evidences of which continue to manifest themselves in the present geopolitical landscape. Thus evidences of the former imperial geopolitical landscape must undoubtedly remain in the present socialist landscape. What is important, in terms of the present examination, is not the fact of their existence, but the question of whether they are of relevance to an understanding of the present. Have the strongholds of the past been entirely abandoned and relegated to a position of quaint and harmless relics, or are they still occupied and used largely for those same purposes for which they were originally constructed?

Having established the importance of bearing in mind this relationship, it must be made clear that it is not the prime purpose here to assess the degree to which the Soviet Union is like or unlike its predecessor, any more than it was the purpose to assess specifically the extent to which Spain was like France, or France was like Austria. The main intention is to consider the extent to which the Soviet Union, in its own right, possesses characteristics which are the same as, or similar to, those of the dominant state. The central question being addressed is the extent to which the Soviet Union conforms to the model of dominance or, alternatively, displays quite different, even possibly unique, characteristics. The geopolitical features of the Soviet Union will now be examined diachronically with this purpose in mind.

As with the move to St. Petersburg some two centuries earlier, the transfer of the capital to Moscow in 1918 was the result both of immediate preoccupations and of longer term considerations. The Treaty of Brest-Litovsk put the fledgling Soviet state into a perilous position, and Moscow, located at the centre of communications, was judged to be a more effective headquarters in the forthcoming war for survival. The

Tsarist capital was maritime, peripheral, cosmopolitan and exposed, while Moscow was central and lay deep in the interior. In the wider perspective, St. Petersburg had been an imperial and aristocratic creation, always closely identified with an alien autocracy while, by contrast, Moscow was Russia's 'soul' situated in the heart of the Russian lands proper. As Kristof observed, the two capitals had been symbolic of the two views as to what Russia should be like. The choice having been made in favour of Moscow, this city was then used by Lenin as 'a nationalistic symbol' to strengthen and legitimate his new regime.[8] Thirty years later at the celebrations for the eight hundredth anniversary of the founding of Moscow, Stalin went out of his way to point to the Muscovite heritage of the Soviet Union. 'By the will of the great Lenin,' he proclaimed, Moscow had once more become the capital and 'the standardbearer of the new Soviet era'.[9]

Russia had been the first nationality to proclaim itself to be a Soviet Republic, and it was around this Russian Soviet Federated Socialist Republic (RSFSR) that the other republics were steadily gathered. By the time of the 1926 census there were six Union republics, these being the RSFSR, Ukraine, Byelorussia, Transcaucasia and the Uzbek and Turkmen Soviet Socialist Republics. The population of the USSR was then 147 millions, and of this the RSFSR alone had 101 millions, nearly 70 per cent of the total.[10] By the time of the 1939 census the number of Union republics had grown to eleven, these being the RSFSR, Ukraine, Byelorussia, Georgia, Armenia, Azerbaijan and, in central Asia, the Turkmen, Kirgiz, Tajik, Kazakh and Uzbek SSRs. By that time the population of the country had risen to 170.5 millions and of this the RSFSR had 109.3 millions, representing a slight decrease to 64 per cent of the total.[11] The RSFSR itself also contained a number of autonomous republics based on smaller nationalities and certain functions, largely concerning their own languages and cultures, were devolved to them.

The whole question of nationality had been a very difficult one for the Bolsheviks to come to terms with. Nations were after all, according to Marx, essentially bourgeois phenomena and their emergence on the scene was bound up with the development of capitalism. They therefore had no natural part to play in the building of communism which was conceived of as arising out of particular socio-economic rather than territorial conditions. The concessions to nationalism which were built into the structure of the Soviet state were consequently of considerable significance and indicative both of the central role of the Russians as custodians of the Holy Grail of Marxist-Leninist purity and of the strength of separatist movements among the other nationalities. In the conditions of war and revolution, even such closely related Slavs as the Ukrainians - the Little Russians - whose history had been linked so closely with that of their Great Russian neighbours, had embarked upon the creation of their own national state. Nationalism, as it had developed in nineteenth century Europe, had been the principal agent in the disintegration of the Ottoman and the Austro-Hungarian Empires and, had it not been for the centralising Bolsheviks, a similar fate might well have been anticipated for the Empire of the Tsars. Thus, at the very outset, there was a dialectical opposition between the territorial and the socio-economic solutions to the legacy involuntarily bequeathed by the Tsars, and this had to be resolved before the new state could function at all. The concession to nationalism was thus a necessity born out of local conditions but it was fitted into an expanded ideological framework. Nationalism, sanitised and radicalised, was taken on board as nothing more menacing than the culture of the common people, the 'narod'. Viewed in this way, nations could be stripped of those bourgeois characteristics ascribed to them by Marx, and become an acceptable expression of the lives of the workers and peasants. Most significantly, from the viewpoint of the state, these nationalities could thus be stripped of any pretensions to real political power and reduced, at least officially, to the level of what Seton-Watson called 'picturesque folklore'.[12] Both Russian pre-eminence and the national cultures would then be subsumed into a new 'Soviet patriotism', and political and economic aspirations could be expressed entirely in Marxist-Leninist terms.

Despite this, in reality the commanding position of the Russians remained a continuing feature of the state. Moscow had the dual role of being the capital both of the RSFSR and of the Soviet Union itself. Here were established the country's most prestigious educational and scientific establishments, its Union ministries and that enormous planning mechanism which was responsible for steering the economic development of the country. While Moscow had not actually been the principal city of the Revolution, following the death of Lenin in 1924 it was soon transformed into the focus of Soviet state iconography, a role which fitted very well with the historic position of the 'Third Rome'. Thus, despite its superficially confederal structure, the reality was one of a highly centralised state with the major decision-making process in most spheres being concentrated in the capital.

Rapid and planned economic development under close state supervision was the principal object of Stalin's Five Year Plans. The development of heavy industry was the central feature of all the early plans, but the geographical distribution of such development was inevitably constrained by the physical conditions within the Soviet Union. As in the countries of the capitalist west, the availability of supplies of the necessary raw materials and energy largely determined the regional distribution of industrial development. It was concentrated into two favoured areas, the Ukraine and western Siberia. In the Ukraine, and in particular in the Donets basin (Donbas), the output of coal and iron was greatly increased. Then in 1930 Stalin decreed that a second coal and metallurgical centre should be established further east. This marked the beginning of the 'Urals-Kuznetsk combine' which linked the Kuznetsk coalfield (Kuzbas) to the rich iron ore deposits in the southern Urals some 1800 kilometres distant. Together with the development of the Karaganda coalfield in northern Kazakhstan, this created a massive heavy industrial region producing huge quantities of coal, iron, steel, heavy metallurgical products and chemicals. By 1940 it already accounted for a significant proportion of the total Soviet output of these commodities and thus made an immense contribution to the Soviet war effort when the regions further west were under German occupation. After World War II the development of the region was further accelerated and by the 1980s it was responsible for over a half of the country's output of coal, iron and steel. Meanwhile other important industrial complexes were being established east of the Urals, in particular in Soviet Central Asia, the Lake Baikal region and the lower Amur basin.

Linked to this impressive industrial expansion was the agricultural development of the regions east of the Urals. This consisted mainly of the planned extension of large-scale cereal production eastwards along an axis from the southern Urals to Lake Baikal. It centred on the latitudinal belt stretching from 50° to 55° north, an area which, since its climate was judged to be not dissimilar to that of similar latitudes in European Russia, had been the traditional line of Russian emigration into Siberia. However, a combination of more adverse physical conditions than had been anticipated, together with insensitive farming methods, made this less successful and contributed to making agriculture for a long period the Achilles' heel of Soviet economic strategy. The development of agriculture further south in Soviet Central Asia proved to be a far more successful venture, in particular the production of such crops as cotton, vegetables and sub-tropical fruit using irrigation waters from the Amu Darya and Syr Darya rivers.

These spectacular economic developments have been accompanied by a rapid increase in the population east of the Urals. By the early 1980s some 80 million people, nearly 30 per cent of the total Soviet population, were living in this area compared to only some 10 per cent at the time of the Revolution. This dramatic change has in part been brought about by the large scale immigration of Russians and other Slavs to central Siberia and the far eastern provinces. However, even more significant is the rapid increase in the Asiatic component of the Soviet population. The combined populations of the five republics of Soviet Central Asia in-

Table 2: The Soviet Union – Population and Population Change by Republic

Republic	Population 1959 (mills)	Population 1986 (mills)	Birth Rate* (per 1000)	Natural* Increase (per 1000)	Percentage Population Increase 1959 – 1986
Armenia	1.8	3.4	22.7	17.2	89
Azerbaijan	3.7	6.7	25.2	18.2	81
Byelorussia	8.0	10.0	16.0	6.1	25
Estonia	1.2	1.5	15.0	2.7	25
Georgia	4.0	5.2	17.7	9.1	30
Kazakhstan	9.3	16.0	23.8	15.8	72
Kirgizia	2.1	4.0	29.6	21.2	90
Latvia	2.1	2.6	14.0	1.3	24
Lithuania	2.7	3.6	15.1	4.4	33
Moldavia	2.9	4.1	20.0	9.8	41
Russian SFSR	117.5	144.0	15.9	4.9	23
Tadjikstan	2.0	4.6	37.0	29.0	130
Turkmen	1.5	3.3	34.3	26.0	120
Ukraine	41.9	51.0	14.8	3.4	22
Uzbekistan	8.1	18.5	35.8	28.4	128
USSR	208.8	278.7	18.3	8.0	33

* denotes figures for 1983

Source: The USSR in Figures (Central Statistical Board of the USSR, Moscow 1986)

creased from 16.7 millions in 1939 to 23 millions in 1959 and then dramatically doubled to 46.4 millions in 1986. During a period when the total Soviet population was itself increasing very rapidly, this still represents an increase from 11 per cent of that total in 1959 to 17 per cent in 1986. The basic reason for this is the extraordinarily high rate of natural increase of the population in that area. As can be seen in Table 2, that of the RSFSR in 1983 was only 4.9 per thousand, while in Kazakhstan it was 15.8, in Turkmen 26 and in Tadjikstan the highest of all at 29 per thousand. While in 1926 the Russian Republic alone had contained 70 per cent of the total Soviet population, by 1939 this had already fallen to 64 per cent. After World War II the proportion continued to diminish, reaching 56 per cent in 1959 and only 52 per cent in 1986. The actual proportion of Russians in the Soviet population in 1986 was 51 per cent. A number of the other Slav republics in the west have been experiencing even more modest rates of natural increase than those of the RSFSR, the Ukraine figure being 3.4 per thousand and Latvia's the lowest of all with a mere 1.3. As a result they too have come to constitute a diminishing proportion of the total Soviet population.

This rapid development of the Asiatic regions of the Soviet Union has been one of the outstanding spatial changes to have taken place within the country since the Revolution. It has been an equalising force, tending to balance out the former marked disparity between European Russia and its vast eastern hinterland. As a consequence, political geographers have identified the emergence of a new centre of power in the east. Hooson referred to it as being 'a new Soviet heartland' and described it as 'a coffin-shaped or crib-shaped axis' stretching eastwards from the Urals to Lake Baikal.[13] With its growing population, improved communications and considerable natural resources this area is seen as being of considerable importance for the future, and indicative of a decided shift eastwards in the geographical centre of Soviet economic power. Siberia and Soviet Central Asia became suppliers of a whole range of raw materials in which the European part of the country had become increasingly deficient. Thus in both war and peace the human and physical resources east of the Urals have been invoked to sustain Soviet world power.

Another significant internal change is the growth which has taken place around the country's peripheries. Notable examples of this are the Leningrad-Lower Neva region, the Baltic Republics, the Black Sea coastlands, Transcaucasia, Soviet Central Asia, Lake Baikal and the lower Amur basin. For the most part these regions are located in the less Russian parts of the Soviet Union[14] and are close to both the external frontiers and either the open sea or large stretches of inland water. Thus the thrust eastwards from the southern Urals to the Kuzbas and beyond which was so important a feature of the earlier Five Year Plans has since World War II been complemented by a new arc of growth stretching around the frontiers of the country from the Baltic to the Sea of Japan. While the historic line of Russia's easterly expansion lay mainly between 50° and 55° north, many of the new growth areas are located well to the south of this, even as far south as 40° north. This development in many ways begins to resemble the American 'sun belt', and it has been associated with a rapid expansion of the country's organic and inorganic resource base.

Yet, despite the strength of these new orientations, Moscow has shown little sign of relinquishing the pre-eminent role assigned to it from the outset. Apart from the very real advantage arising from its location, it has been sustained by the centralising role of the Communist Party of the Soviet Union and its preponderant position in the nation's policy-making structures. The capital's position since World War II has also been reinforced by its role as the centre of the Soviet information media, and of the comprehensive internal air transport network. The powerful centripetal structure focusing on Moscow is thus at variance with the pronounced centrifugal tendencies in population distribution, economic growth and the divergent rates of increase of the various component nationalities. This situation represents the geographical manifestation of that dichotomy between the territorial and the socio-economic condition which, as has been observed, had been present since the establishment of the Soviet state and upon which its internal spatial structures are founded. While the principal thrust of the socio-economic thinking is towards a centralisation of decision-making, greater overall uniformity, the minimisation of territorial variation and the conquest of nature, that of the territorial is towards localism, greater fragmentation and the encouragement of political, economic and cultural variation. While Marxist socio-economic doctrines remain the official orthodoxy, the state has been forced to make concessions in the interests of the maintenance of relative internal harmony.

Another aspect to be considered is the relationship of the actual territorial extent of the Soviet Union to that of the Russian Empire. This relationship has three principal geopolitical aspects. Firstly, there was the regaining of those possessions of the former Empire which had been lost to neighbouring states in the wake of World War I. In both the centre and the north this re-acquisition stopped short of a complete resumption of the imperial frontiers. In the centre, the move forward was approximately to the Curzon Line which had been accepted in principle by the Soviets but not by the Poles. Poland itself was, however, not reincorporated and was even given German territory as compensation for what had been lost. In like manner, while the Baltic-Gulf of Finland coastlands were reincorporated, Finland itself, west of the strategic Lake Ladoga-Gulf of Finland isthmus, retained its independence. Poland and Finland had together been the most recent acquisitions in the west, dating from the victory over Napoleonic France, but the Soviet Union refrained from moving forward to encompass the full extent of this former imperial territory. The post-World War II frontiers did however result in the incorporation of the whole of historic Kievan Russ and the enlargement of the Baltic 'peep-hole' into the post-Petrine 'window on the west'.

Secondly there was the expansion beyond any previously held frontiers into completely new territory. As has been seen, this took place in only two relatively small areas, namely East Prussia and Galicia-Ruthenia. In both cases, modest territorial acquisitions produced important strategic gains. The acquisition of the northern part of East Prussia gave the Soviet Union the major port of Kaliningrad (Königsberg) and the new naval base of Baltiiysk (Pillau) on the southern coast of the Baltic. The acquisition of Galicia-Ruthenia moved the Soviet frontier for the first time west of the strategic Carpathian mountain barrier and put her into a far stronger military position for controlling the middle Danube.

Thirdly there was the extension of Soviet control into the lands beyond her western frontiers after World War II. This acted as both a military buffer zone in the classic manner and as a sphere of ideological and economic domination. More specifically, it had important territorial consequences for the countries lying around the Soviet Union's western frontiers. The Soviets achieved hegemony over a substantial slice of the defeated hegemonial power of Mitteleuropa, including almost the whole of its historic core region of Brandenburg-Prussia. In addition to this they also achieved hegemony over most of the middle and lower basins of the Danube and of the North European Plain westwards to the Elbe. In economic terms the transnational sub-Carpathian industrial region which extended across parts of East Germany, southern Poland and northern Czechoslovakia and the rich agricultural lands of the Pannonian basin also came within their sphere. Thus in the aftermath of World War II the Soviet Union both improved upon the former frontiers of Imperial Russia and established for the first time a buffer zone stretching deep into the heart of Europe.

The geopolitical characteristics of the Soviet Union will now be considered in the context of those of the dominant state, and an assessment will be made of the extent to which there is conformity with the normative model based upon them. This will then be assessed against the four possible explanations of the nature of the geopolitical structure of the Soviet Union which have already been pointed out.

The first act of the new Soviet government to have fundamental geopolitical implications was the transfer of the capital from Petrograd to Moscow. This represented a return of the centre of political power, after an absence of two centuries, back to the Muscovite core region in the geographical heart and spiritual soul of the country. It was also a return to the centre of European Russia's internal communications adjacent to the Valdai elevation and at the pivot of the country's rail network. As they had done earlier from the same strategic location, the Russkii then gathered the other East Slavs, the Rossiiskii, around them and subsequently went on to add the non-Slav populations of Transcaucasia and the Baltic, who had also been former subjects of the Tsar. The leading role of the Russian core nation was maintained and reinforced, and the capital remained within its historic core region. This special position, as Niebuhr pointed out, is not itself based on any dogma, any more than was the special position of Rome as the source of authority for early Christianity. In both cases it was the contingencies of history which supplied the source of authority, and the role of dogma was to supply the necessity for there to be a single source of authority. 'Russia', said Niebuhr, 'was not designed to rule the kingdom of God on earth according to the Marxist creed', but rather was forced into the position of assuming 'hegemony of the cohorts of socialism in achieving the victory of the righteous over the unrighteous imperialists'.[15] A fundamental geopolitical characteristic of the Soviet Union is thus the primacy of Russia, and this is a position directly inherited from its predecessor.

The immense diversity which was also a feature of the new state was most strongly in evidence in the peripheral regions of the west and south (Figure 6.3). The implementation of a series of measures designed to reduce this diversity was legitimised through recourse to the official state ideology and reinforced by the powerful political iconography which centred on the capital. The expansion out from the Muscovite-Russian core region was rapidly to encompass most of the territory of the former Empire, but while the deliberate policy of Russification pursued by the later Tsars was officially rejected in favour of one of 'self-determination', most of the nationalities were in fact forcibly reincorporated. The justification put forward was that the proletariat of these nationalities desired unification with the triumphant Russian proletariat and that only 'unrepresentative bourgeois and feudal cliques' really sought independence.[16] In diachronic geopolitical terms the Soviets thus re-enacted within a decade the whole process of expansion from Muscovy to Russian Empire which had originally taken four centuries to accomplish. The establishment east of the Urals of a new centre of economic power was the extension into the industrial sphere of the former political and commercial controls.

Figure 6.3: Geopolitical Characteristics of the Soviet Union

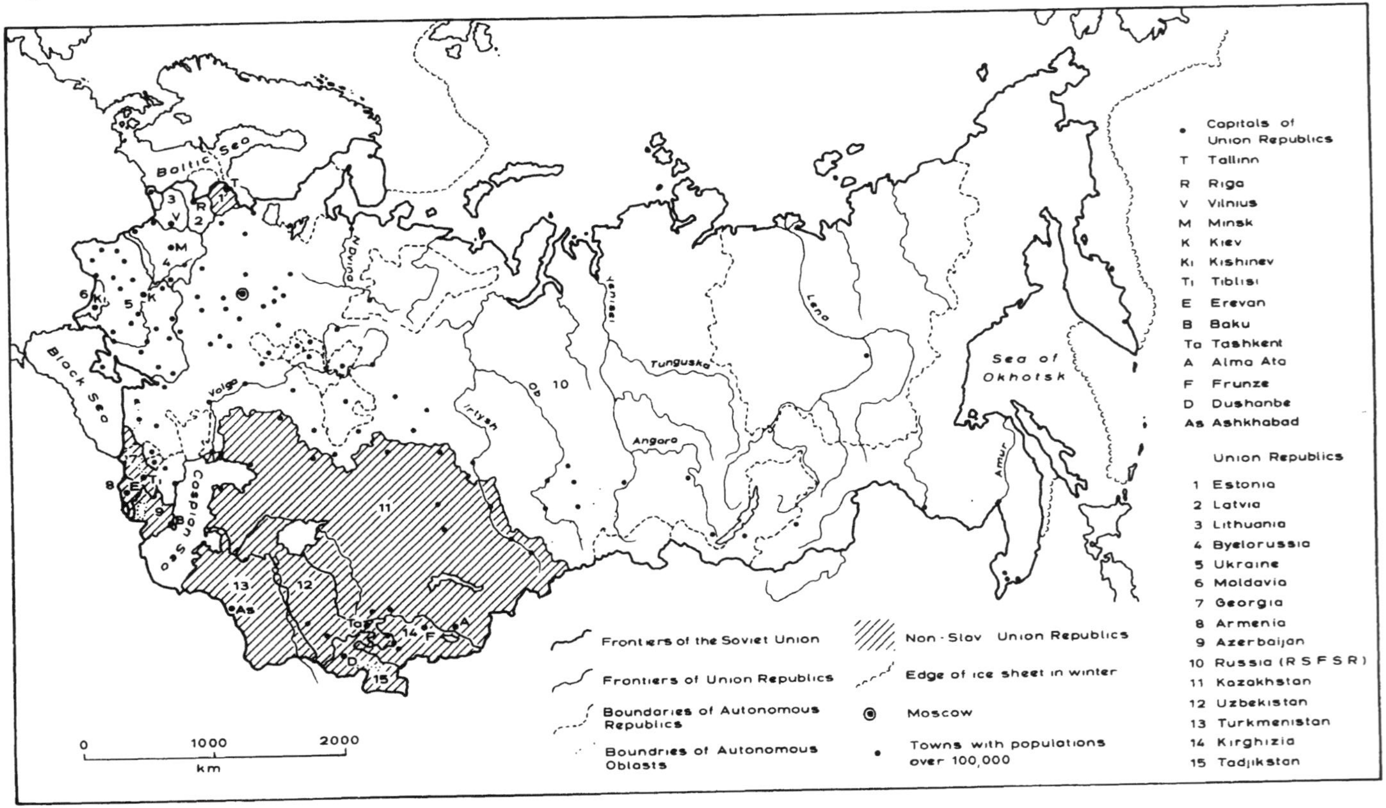
Baltic Sea
Black Sea
Caspian Sea
Sea of Okhotsk
N Dvina
Volga
Irtysh
Ob
Yenisei
Tunguska
Angara
Lena
Amur
0 1000 2000
km
Frontiers of the Soviet Union
Frontiers of Union Republics
Boundaries of Autonomous Republics
Boundries of Autonomous Oblasts
Non-Slav Union Republics
Edge of ice sheet in winter
Moscow
Towns with populations over 100,000
Capitals of Union Republics
T Tallinn
R Riga
V Vilnius
M Minsk
K Kiev
Ki Kishinev
Ti Tiblisi
E Erevan
B Baku
Ta Tashkent
A Alma Ata
F Frunze
D Dushanbe
As Ashkhabad
Union Republics
1 Estonia
2 Latvia
3 Lithuania
4 Byelorussia
5 Ukraine
6 Moldavia
7 Georgia
8 Armenia
9 Azerbaijan
10 Russia (R S F S R)
11 Kazakhstan
12 Uzbekistan
13 Turkmenistan
14 Kirghizia
15 Tadjikstan

   The second phase of Soviet territorial expansion, that outwards from the 1923 frontiers, was triggered by external forces, namely the attack by the hegemonial power of Mitteleuropa making its final bid for dominance over the whole of the North European Plain. The Soviet territorial expansion which followed the defeat of Germany resulted in the frontiers being moved forward to the Baltic-Black Sea isthmus. Since Poland remained outside these new frontiers, they proximated more closely to the shortest line across the isthmus - that from the Gulf of Klaypeda to the estuary of the Dniestr - than the former imperial frontiers ever did. Since over much of their extent they also followed rivers, in particular the Prut and the Bug, they came as close to being western 'limites naturelles' as the uninterrupted North European Plain was ever likely to make possible. At the same time the Soviet Union acquired a substantial coastline on the Baltic Sea. Outside the isthmian frontier stretched that wide buffer zone which had the effect of projecting Soviet military power right into the heart of Mitteleuropa itself (Figure 6.2).
   On the basis of the characteristics which have been examined, the Soviet Union bears a close resemblance to the Russian Empire. Nevertheless, there are a number of significant differences between the two, the most important concerning their core regions and frontiers. The characteristics of the original and historic core regions used in the model are the result of a long period of development, which is something a relatively young state like the Soviet Union ipso facto can not have had. In the particular circumstances of the birth of the Soviet state, the historic core was a 'given', a ready-made centre of power to which it gravitated and from which it could most effectively begin to exercise political control. Yet there were marked historical parallels between the use of the core region by the Soviets and its role in the early Russian state. For instance, Moscow's role as a defensive base by the 'Reds' against the 'Whites' and the Interventionist powers certainly had a parallel in the historic role of Muscovy as a stronghold against the Mongols, the Poles and the Lithuanians. Then again during World War II the historic core took on the role of an embattled fortress, the rallying place for resistance against the aggressor. The Soviets had not themselves been responsible for the construction of the fortress, but they inherited it and used it largely for its original purpose. Leningrad, the Petrine maritime core, was also converted into a defensive fortress during the Great Patriotic War, a role for which it, unlike Moscow, had little historic precedent. However, this subsequently had the important effect of ending its position as an alien and artificial legacy of the Tsarist Empire and confirming its venerated role as the city of Revolution. As to the old economic core in the Ukraine, this considerably diminished in relative importance as a result of the developments which

took place east of the Urals. Again however, although constituting a massive internal spatial change, the eastward shift of the centre of industrial power did not represent a fundamental divergence from the inherited imperial structure. Rather it was the supplementing of one economic core region with another which possessed a much greater resource potential.

In respect of the situation on the Soviet frontiers, the establishment of a buffer zone has no real parallel in the past, although the influence of the Russian Empire had reached, partially and intermittently, as far as the Balkan peninsula. The projection of effective control to the Elbe split Mitteleuropa into two and thus secured the base from which the twentieth century invasions of Russian and Soviet territory had been mounted. This was well in excess of anything which the Russian Empire had been able to achieve, although Marx himself had referred to the Stettin-Trieste line with considerable prescience as 'the natural frontier of Russia'.[17] The creation of a buffer zone of these dimensions is by no means a departure from the model. The existence of satellite states having a measure of autonomy has invariably been a feature of the peripheries of the dominant state. In such peripheral areas political, economic and cultural diversity has generally made it prudent for the dominant state to adopt a more flexible and subtle approach to the exercise of control.

If, for the present, those characteristics pertaining to the actual condition of dominance are excluded, then the Soviet Union possesses some 84 per cent of those characteristics upon which the model of the dominant state is based. In reaching this figure, the core regions have not been considered as being specifically Soviet characteristics, but as being geopolitical survivals from the former state. However, there is a case for including them as being positive Soviet characteristics since, as has been observed, they have actually fulfilled a role in Soviet development analogous to that of the cores of the dominant states previously examined. If they are included, then the Soviet Union achieves almost complete conformity with the model. However, in terms of the requirements of Stage 3 of the model, specifying control over the macrocore of the culture area and the greater part of its territory, the Soviet Union has not on the face of it achieved the undisputed position of a dominant state. In assessing the extent of the attainment of the Soviet Union in this respect certain specific territorial questions need to be further examined.

The civilisation to which Russia belongs has been traced back through Muscovy via Kievan Russ to that of Byzantium, the macrocore of which was located around the shores of the Bosphorus. Russia's historic expansion towards this focal region was motivated by a variety of commercial, cultural and strategic considerations, but central to them was the perception of the region as being a Christian 'terra irredenta'. After the Revolution official claims to this territory were renounced,[18] but just as the Tsars and the Pan-Slavs had been drawn towards it for their own particular reasons, so also were their Soviet successors. Those historic motivations deriving from Orthodoxy may have given place to new ones based on raisons d'état, but acquisition of the Dardanelles remained both a strategic objective and a national aspiration.[19] In the post-Versailles era its attainment was an extremely remote prospect, but after World War II the situation was a very different one. However, while the victorious Soviets thrust westwards into the heart of Europe, southward movement was far more limited and hesitant, though for a time both Yugoslavia and Albania were within the Soviet sphere, giving the Soviet Union better access to the Mediterranean than the Russian Empire had ever had. Any moves towards the south, whether by the Soviet Union or by other communist states, were however soon countered by the build-up of Anglo-American support for Greece and Turkey. Soviet demands made at the time of the German capitulation for the Dardanelles together with a slice of Turkish territory met with a cold response from the Western powers.[20] The response to the Soviet demand for Libya, a former Italian colony, subsequently made at the Potsdam Conference was equally unfavourable.[21] Ernest Bevin, the British Foreign Secretary, came to fear that the Iron Curtain was in danger of being drawn southwards into the Mediterranean, thus cutting British communications with her possessions in the east.[22]

While Orthodoxy had emanated from Byzantium, the state ideology which replaced it had its origins in northern Europe. The triumph of Marxism can be regarded as having been a final consequence of the country's post-Petrine northern orientation. It brought with it the major problem that while the Marxist perspective on the relationship of capitalism to socialism was intended to be applied to the more advanced societies, the state inherited by the Soviets was, in the words of one economist, 'more akin to a backward primary-producing country than an advanced industrial one'.[23] Since the new Soviet state was intended partly for the purpose of emulating western material advance, special measures were needed to fit the doctrine to local conditions. The politico-economic structures then set up to accomplish this presented a stark contrast to those which had been evolving over a long period in northern Europe. They were characterised by centralisation, uniformity and rigid controls, all operating within the framework of a world-view imposed by the state. These characteristics have been identified as being those of states which have aspired to a position of dominance. The politico-economic philosophy which was developed by the

Bolsheviks was, in fact, the product of the fusion of two highly dissimilar strands of thought. It derived from a synthesis of elements of nineteenth century European political philosophy with their antithesis, a pre-existing system impregnated with Byzantine and oriental features. The fusion of these, taking place within the particular geographical and historical conditions of the Soviet Union, produced something very different from either of them. The system which emerged was officially labelled Marxist-Leninist, but it could more appropriately be thought of as being Leninist-Stalinist, since in reality it departed from most things with which Marx could have been expected to identify. To Arendt it was more like a new religious theory, founded in Russia's position as 'the true divine people of modern times'.[24] To Wittfogel it was based on 'a quasireligious vision',[25] while Toynbee thought of it as being an emotional and intellectual substitute for Orthodoxy.[26] What resulted, said Berdyaev, was, despite its universalist ideology, a deeply Russian phenomenon.[27]

The real question, according to Kennan, was not how much Bolshevism (sic) changed Russia, but how Russia, 'the natural environment, the Asiatic frontier, the Black Sea civilisation, the Byzantine Church, the backwardness' in the end changed Bolshevism.[28] Kennan's question was largely rhetorical, but his answer had to be that the political philosophy imported from Europe had been transformed out of all recognition by the geographical conditions of the area into which it had been transposed. As Trotsky, the early Menshevik, saw it, that democratic urge inherent in the aspirations of the Soviets came to be replaced by centralisation and rigidity as a direct result of the tight control which Moscow succeeded in imposing. This led to the seeming paradox that 'the most Occidentalised of Russian revolutionary ideologies resulted...in the de-westernisation of Russia'.[29] Whatever the other consequences of this may have been, it certainly had the effect of establishing, for the first time, a truly autonomous system in the Russian lands. This was geopolitically more significant than had been the dubious and politically motivated 'Third Rome' concept which in reality had never amounted to much more than the transposition of certain attributes of a moribund culture from the Bosphorus to the Mezhdurechie. Then Byzantium-on-Dniepr had been transformed into Byzantium-on-Moskva, and the legacy of the Mongol conquests had certainly not produced attributes which one would have associated with a 'Third Rome'. Given its high measure of conformity with the political characteristics of the dominant state, together with the development of cultural autonomy, Bolshevik Moscow was perhaps a more genuinely 'new' Rome than its medieval predecessor ever had been. Despite the eschatological pronouncements of the Abbot Philotheus,[30] it might be more apt to consider it as being an emerging 'Fourth Rome' since its lineage can be traced back dialectically both to the 'imperium orbis terrarum' of Rome itself and to the northern European commercial-industrial complex which was the motor of nineteenth century western imperialism. As a consequence, in place of the quasi-colonial status of Kiev, Muscovy and even St. Petersburg, the creation of what was, in effect, an autonomous cultural macrocore was embarked upon in Stalin's Russia. Through the implementation of a xenophobic policy of isolation and insulation from the western world-economy, the country's autonomy was rapidly extended from the political into the economic, cultural and ideological spheres. The centre of the new macrocore was Russia proper, and its 'Holy Places' were Leningrad - the Medina of the Revolution - and Moscow - the Mecca of the new socio-economic creed. 'Muscovite eschatological messianism', said Sarkisyanz, 'sparked by imported Marxism, exploded into the visionary message of world revolution'.[31]

The identification of an emerging macrocore within its territory then raises the question of whether the Soviet Union ipso facto became a dominant state in the western ecumene. In terms of its geopolitical characteristics in the aftermath of World War II, the Soviet Union undoubtedly attained a position very similar to those of the dominant states of the past. However, although become the largest and most powerful state in the western ecumene, it had clearly not become undisputedly its dominant state. Although its influence extended deep into the centre of Europe, the maritime crescent stretching from Scandinavia to the eastern Mediterranean, bolstered by American power, remained firmly outside its grasp. If it is to be considered as being a dominant state, and it had acquired many of the geopolitical characteristics of one, then this had come about in a Eurasian rather than a specifically European geographical context. The geopolitical implications of the emergence of an indigenous macrocore were that Leninist-Stalinist Russia had in effect withdrawn from participation in the wider European culture area and instead had closed the frontiers and turned inwards to establish an alternative cultural centre of its own. While this position was in part at least forced upon the Soviet Union as a consequence of its post-Revolutionary international isolation, the reorientation away from Europe dated back to the Tsarist period. At the time of the collapse of the Empire, Siberia and the far eastern provinces were already growing in importance. The Revolution then had the effect of accelerating this tendency and giving it a revived raison d'être through the substitution of a fresh and, at least to start with, credible state ideology for one which was old and discredited. Eurasianism - Evraziistvo - was thus both a geographical product of the conditions prevailing after the Revolution and also a culmination of Russia's old 'eastern dream'.[32] It was the result both of disillusion with Europe and of a realisation of the opportunities offered by the Asiatic connections which the new state had inherited. The Eurasian concept, according to Kristof, led 'from the idea of a symbiotic Slav-Asiatic "Middle World" allied with "Asia properly speaking" against Europe'.[33] It proclaimed, in Berdyaev's words, 'light from the East which is destined to enlighten the bourgeois darkness of the West'.[34] The Pan-Slav ideas of the nineteenth century were seen as having been far too restrictive in scope, and Evraziistvo came down much more firmly in the older Russian messianic tradition, now taking the form of the rallying of mankind against the hegemony of Europe.

Viewed from this Eurasian perspective, the Soviet Union fits the model of the dominant state more closely than the Russian Empire does. The vast territory lying between the Baltic-Black Sea isthmus in the west and the Sea of Okhotsk in the east and between the Arctic Ocean and the mountain vastnesses of central Asia took on the appearance of a 'natural' unit. It was the geographical space occupied by the 'Middle World' destined to be developed under the guidance of Russia according to the received wisdom of the state ideology. The Soviets, as Kristof saw it, had inherited a cultural-political domain and along with it they had acquired 'a certain teleological impetus conditioned by history and geography'.[35] Since Moscow had been the city from which, as one nineteenth century writer put it, 'the hand of Russia was laid upon the east' thus acquiring 'entire kingdoms of most valuable and undetachable lands', the transfer of the capital there could also indicate a statement of the will to build a Eurasian rather than a European future. 'The preservation of iconography', said Gottmann, 'could also dictate the closing of national space'.[36]

While developments within the Soviet Union can thus be interpreted more logically within a Eurasian rather than a European framework, the movement towards greater unity among the states of the European maritime crescent was motivated in large part by mistrust of Soviet motives in the western ecumene. The progress towards unity was initiated by the states of the Lotharingian axis and its successful development then centred on the north-west European macro-core.[37] The official Soviet view of maritime Europe is that it is of diminishing significance. As a consequence of the continued inadequacy of its socio-economic system, the iron laws of historical development prescribe that capitalist Europe is doomed to extinction. According to this view, Europe is a phenomenon of the past, 'an atrophied body, merely an appendix of Eurasia'. Its collapse being inevitable, no special action was therefore called for to bring it about. The principal role of the Soviet Union was thus to help guide the process and to prepare to assume that 'hegemony of the cohorts of socialism' in the struggle with the 'unrighteous imperialists'. Consciously or unconsciously it was envisaged that the cohorts would be 'gathered' in the historic Muscovite fashion. This had already taken place right in the heart of Mitteleuropa itself and one might imagine, in Kristof's words, 'a greater Eurasia rolling ... down to the Stettin-Trieste line, perhaps to the very shores of the Atlantic'.[38] The political fragmentation of the western ecumene, which had been the most persistent feature of its recent history, would then be finally ended by the merger of a socialist Europe into a greater Asia. In this way Europe, already reduced to a maritime crescent fringing the Soviet-dominated continental core, would cease to exist as an independent geopolitical entity and be merged into a new Eurasian entity. The geographical focus of such a Eurasia is the great plain stretching eastwards from northern Europe to the borders of China, and Russia has been historically its central organising power.

The characteristics of the geopolitical structure of the Soviet Union, set in the context of the Marxist-Leninist historical thesis, lead one to the conclusion that by the 1950s the Soviet Union had already almost attained a position of dominance in the western ecumene at least as great as had the Ottomans in the sixteenth century. Contingent upon the proposition that the principal cultural macrocore of the western ecumene was in course of eastern displacement in much the same manner as it had been displaced from the western Mediterranean to north-west Europe in the seventeenth century, the Soviet Union was all set to assume the position of the universal state of the western ecumene. However, even if this had been a tenable view in the early 1950s, it had certainly ceased to be so a decade later. By then serious discrepancies had opened up between the actual position of the Soviet Union and the model of the dominant state. The most significant of these concerned the position of the frontiers, the macrocore and the internal power structure.

The security buffer around the Soviet frontiers remained incomplete. This was especially so in Asia where the end of Eurasian communist solidarity, under a joint Soviet-Chinese hegemony, made the newly-developing peripheral regions highly vulnerable. As to the western ecumene itself, the buffer zone would most certainly, given favourable international conditions, have been extended southwards towards the Dardanelles and Anatolia. As has been observed, the region had historically been an important element in Russian spatial ideology, and it continued to influence the Soviet perception of its optimum military and economic sphere. After World War II it thus remained both a denial and a danger, a geopolitical if no longer a religious 'terra irredenta', firmly controlled by a power which, despite its neutrality during World War II, was traditionally hostile. Even more disquieting from a Soviet perspective was the fact that it was politically in the western camp and was in course of attempting to integrate into the new European international system.

Although the Soviet Union had become the most powerful state in the western ecumene, the other erstwhile great powers of Europe not only remained firmly outside its sphere of control, but were also themselves engaged in greater mutual co-operation than had taken place on the continent for many centuries. With the north-west European macrocore as their motor, they gained greater success than did those areas lying within the Soviet sphere. The Soviet zone of Germany without Berlin, in the view of William Manchester, was no more than 'one big farm' while the western zones contained the massive Rhine-Ruhr industrial complex which had been the 'anvil of the Reich'.[39] Manchester was certainly exaggerating to some extent, but it is valid to make the point that, although they had conquered territory, the Soviets were far from being the winners in economic terms. Their buffer zone consisted largely of the poorer parts of Europe, and was certainly underdeveloped by western standards. This might in itself have been of less consequence had the Marxist-Leninist historical scenario become a reality, but the growing gap between theory and fact was becoming all too apparent in the political geography of post-war Europe. Far from being in the course of the predicted 'inevitable' decline, the north-west European macrocore had gained a new lease of life under a modified version of the capitalist system. The era of the national 'economic miracles' of the 1950s laid the foundation for the greater 'European miracle' of the 1960s in which states so recently at war with one another agreed to bury the animosities of the past and come together in the wider interests of them all. This represented a spectacular recovery from the moral degradation and physical devastation produced jointly by Fascism and war. While this was taking place in the west the Soviet Russian macrocore, putative heir to that of north-west Europe, was showing disturbing signs of weakness. In respect of industrial strength, economic growth rates, scientific and technical innovation, cultural vibrancy and ideological dominance, it was by the 1960s definitely under-achieving. The stridency of Stalin's immediate heirs and their continued xenophobia undoubtedly arose from the realisation that the country's international position, both inside and outside the Soviet sphere, was a deteriorating one.

Finally, there are the implications of the spatial changes which have taken place both within the territory of the Soviet Union and in its sphere of influence. The most important internal spatial changes since World War II have been the moving of the centre of gravity eastwards, the development of a number of peripheral regions and the greatly increased importance, both in demographic and economic terms, of the non-Russian nationalities. The best interests of many of these nationalities and regions cannot necessarily be assumed to coincide exactly with those of Russia itself. The dominant position of the Soviet Russian macrocore is bound to become a less automatic one in a diverse Eurasian state than in a state which was essentially heir to the Rossiiskaia Imperiia. The automatic position of the Rossiiskii as the hegemonial nation thus becomes more open to question than at any time since the Polish-Lithuanian bid for supremacy was defeated by Muscovite Russia. This 'centrifugal impulse' can be observed still more strongly in the western buffer zone where, in Liska's words, 'the natural recoil of smaller powers from a regionally preponderant great power'[40] has been greatly reinforced by the attractions of the west. Looking outwards in this way, many of the countries of the zone have come to question the role imposed upon them in the 1940s as daughter nations in a 'socialist commonwealth' gathered, like those within the Soviet Union itself, around the all-pervasive Matushka Rossiia.

In conclusion, the extent of the completeness of Soviet dominance, even as an evolving and developing phenomenon within both Marxist and geopolitical perceptions, has been seen to be very much open to question. In this respect the Soviet Union is again like, rather than unlike, the dominant states of the past, since all the aspirants to a dominant position in the western ecumene in modern times have had to face the fact that the extent of their control within it, although overwhelming, was never a total one. Dominant they may have been, but not one of them succeeded in achieving the ambition of establishing an undisputed imperium over the whole of the western ecumene. Their response to this situation of incompleteness was two-fold. First, there was an attempt to impose greater uniformity within the territory already controlled, and second an attempt to extend still further the extent of that control. The overall result was 'to encompass more and more geography' and by so doing gain 'more and more problems'.[41]

Should the Soviet Union choose to respond in like manner to the increasing inadequacies of its position in relation to the western ecumene, then one might have to anticipate an attempt at some stage to impose tighter controls within its sphere of influence, and possibly also further attempts at territorial aggrandisement. The high degree of Soviet conformity with the model of the dominant state indicates this as being a possible, even a probable, policy option. However, there are also geopolitical characteristics which could have the effect of modifying or even of altering the country's behaviour patterns. In this context the extent of the 'uniqueness' of the Soviet Union as a geopolitical phenomenon will be assessed. If it does turn out to possess certain unique characteristics, then this must inevitably have important implications for the position of the Soviet Union in any interpretation of the world international scene.

NOTES AND REFERENCES

1. Named after the British Foreign Secretary, Lord Curzon, the line extended from Grodno via Brest-Litovsk and Przemysl to the Carpathian mountains. It had been proposed to the Polish government by the British Prime Minister, Lloyd George, in July 1920.

2. M. Pap, 'The Ukrainian Problem' in W. Gurian (ed.), *Soviet Imperialism: Its Origins and Tactics* (University of Notre Dame Press, Notre Dame, Indiana, 1953), p. 43.

3. G. Liska, *Russia and World Order* (Johns Hopkins Press, Baltimore, 1980), p. 66.

4. The Treaty of Riga between Poland and Russia, October 1920. See I. Bowman, *The New World* (Harrap, London, 1922), pp. 333-7.

5. S. deR. Diettrich, 'Some Geographic Aspects of the Russian Expansion', *Education*, Feb. 1952, p.9.

6. The term Iron Curtain was popularised after it had been used by Winston Churchill in a speech delivered at Fulton, Missouri, in March 1946. However, it had actually been used earlier by Joseph Goebbels, Nazi propoganda minister, and by Count Schwerin von Krosigk, who became a minister under Doenitz in the last government of the Third Reich. Both appear to have foreseen, with a clarity perhaps stemming from the German geopolitical tradition, the impending division of Europe into two hostile camps.

It has been suggested that the term dates back even further still, and was used shortly after the Bolshevik Revolution. See *Oxford Dictionary of Quotations*, 2nd ed., 1962, p. xix, Corrigenda 144:15.

7. The German Democratic Republic (DDR) established under Soviet control and supervision in 1949 continued to be referred to as '*der Sowietisches Zone*' on West German maps until after the signature of the Treaty of Friendship between the Federal Republic and the DDR in 1972.

8. J. Gottmann, 'Organising and Reorganising Space', in J. Gottmann (ed.), *Centre and Periphery: Spatial Variation in Politics* (Sage, London, 1980), p. 222.

9. F.C. Barghoorn, 'The Image of Russia in Soviet Propaganda' in W. Gurian (ed.), *Soviet Imperialism: Its Origin and Tactics* (University of Notre Dame Press, Notre Dame, Indiana, 1953). p. 154. The quotation is from the Stalin Proclamation of 7 Sept. 1947.

10. *USSR Information in Figures, 1917-1927* (Ministry of Information, Moscow, 1928).

11. *Census of the USSR* (Ministry of Information, Moscow, 1940).

12. H. Seton-Watson, *Nations and States. An Enquiry into the Origins of Nations and the Politics of Nationalism* (Methuen, London, 1973), p.18.

13. D.J.M. Hooson, *A New Soviet Heartland?* (Van

Nostrand, Princeton, New Jersey, 1964).

14. Over a long period, Russians have also been colonising these peripheral regions. The Lake Baikal region has been colonised by Russians since the middle of the seventeenth century, and they have displaced large populations of native peoples. As has been observed, St. Petersburg was for long regarded as being a foreign city. Since 1917 it has been regarded as the city of the Revolution, although it retains much of its Baltic quality. Elsewhere since the Revolution large numbers of Russians have settled around the peripheries of the country. There are over one million non-nationals in the small populations of both Latvia and Lithuania, and most of these are Russians. Similarly there are 1.5 million non-nationals in the Kirgizian Republic and 4 millions in the Uzbek. (Soviet Census data).

15. R. Niebuhr, *Nations and Empires* (Faber, London, 1959), pp. 247-8.

16. H. Seton-Watson, 'The Historical Roots' in C. Keeble (ed.), *The Soviet State. The Domestic Roots of Soviet Foreign Policy* (Gower for the Royal Institute of International Affairs, London, 1985), p. 16.

17. K. Marx, in K. Marx and F. Engels, *The Russian Menace to Europe* P.W. Blackstock and B.F. Hoselitz, (eds.), (Allen & Unwin, London, 1953), p. 132. Originally in *New York Tribune*, 12 April, 1853.

18. T. Stoianovich, 'Russian Domination in the Balkans' in T. Hunczak (ed.), *Russian Imperialism from Ivan the Great to the Revolution* (Rutgers University Press, New Brunswick, New Jersey, 1974), p. 230.

19. W. Kolarz, *Myths and Realities in Eastern Europe* (Lindsay Drummond, London, 1946), p. 84.

20. T.H. White, *Fire in the Ashes: Europe in Mid-Century* (Cassell, London, 1954), pp. 335-40.

21. M. Wight, *Power Politics* (Royal Institute of International Affairs and Penguin, Harmondsworth, 1986), p. 146.

22. A. Bullock, *Ernest Bevin, Foreign Secretary 1945-1951* (Heinemann, London, 1983), pp. 132-3.

23. M. E. Falkus, *The Industrialisation of Russia* (Macmillan, London, 1986), p. 5.

24. H. Arendt, *Imperialism* (Harcourt Brace Jovanovich, New York, 1968), p. 113.

25. K. A. Wittfogel, *Oriental Despotism. A Comparative Study of Total Power* (Yale University Press, New Haven, 1957), p. 440.

26. A.J. Toynbee, *A Study of History*, abridged D. C. Somervell (Oxford University Press, 1946), p. 204.

27. N. Berdyaev, *The Russian Idea*, trans. R. M. French (Geoffrey Bles, London, 1947), p. 250.

28. Quoted by D. Yergin, *Shattered Peace: The Origins of the Cold War and the National Security State* (André Deutsch, London, 1978), p. 37.

29. E. Sarkisyanz, 'Russian Imperialism Reconsidered' in T. Hunczak (ed.), *Russian Imperialism from Ivan the Great to the Revolution* (Rutgers University Press, New Brunswick, New Jersey, 1974), p. 59.

30. See reference 11, Chapter 5.

31. E. Sarkisyanz, 'Russian Imperialism Reconsidered', p. 58.

32. L.K.D. Kristof, 'The Russian Image of Russia' in C.A. Fisher (ed.), *Essays in Political Geography* (Methuen, London, 1968), p. 370.

33. L.K.D. Kristof, ibid., p. 373.

34. N. Berdyaev, *The Russian Idea*, p. 350.

35. L.K.D. Kristof, 'The State-Idea, the National Idea and the Image of the Fatherland', *Orbis*, vol.II, 1, Spring 1967, p. 252. The reference is to P.Kh. Grabbe in his *Memoirs*, January, 1854.

36. J. Gottmann, 'Organising and Reorganising Space' in J. Gottmann (ed.), *Centre and Periphery. Spatial Variation in Politics* (Sage, London, 1980), p. 222.

37. G. Parker, *A Political Geography of Community Europe* (Butterworths, London, 1983), Chapter 1.

38. L.K.D. Kristof, 'The Russian Image of Russia', p. 383.

39. W. Manchester, *The Arms of Krupp* (Bantam Books, London, 1970), pp. 723-4.

40. G. Liska, *Russia and World Order*, p. 119.

41. D. Yergin, *Shattered Peace*, p. 196.

SEVEN

THE SILENT CASTLE : A CASE OF GEOPOLITICAL
UNIQUENESS?

In general, western attitudes to the Soviet Union have
accorded with the Soviet contention that the country is a
unique political phenomenon. Reinhold Niebuhr expressed the
view that 'The world is confronted with what seems to be an
entirely unique emergence in history.'[1] However, while the
Soviets, along with those who subscribe to their ideological
world-view, have perceived this uniqueness in idealistic
terms, opponents or sceptics, whether consciously or unconsciously, have more often used geopolitical arguments to
reinforce their contentions. From this standpoint the Soviet
Union emerges as a uniquely dangerous, subversive, expansionist and, at worst, evil force in the world. This view rests
on two related elements in the Soviet scene: the proselytising
nature of Marxist-Leninist ideology and the historical expansionism of Russia, now metamorphosed into the Soviet Union.
The two have been fused into the perception of a political
unit which displays a set of interrelated geographical, historical and ideological attributes. Such attributes were
grouped together into what Serge de Chessin proclaimed in
1930 to be 'Darkness from the East'.[2] According to him,
'The "Red" doctrine marks a set-back and a paralysis of
progress ... It really embodies the worst form of reaction
(and) represents the worst form of stagnation, a petrification
of thought. It is no longer light that comes from the East; it
is night, dense night, total darkness, a new Mongol invasion,
but on a spiritual plain this time. The Apocalyptic horsemen
advance from a Bolshevised Asia.'[3] De Chessin concludes
that "le grand soir" surely needs no cosmological justification',[4] since it indicated that Russia was now engaged in
'summoning the Scythians to the place of the Varangians'.[5]
In saying this, de Chessin places the phenomenon clearly into
the Russian geopolitical context. It was an expression, he
said, of the country's experience and thus 'Lenin belongs to
the line of Gengis Khan and Tamerlaine. Social revolutions
reduce themselves in his mind to the idea of invasion.' Thus
instead of the cosmological comes the geographical expla-

nation. 'The Eurasiatic steppes, that immense shapeless plain stretching between two worlds' had always, as de Chessin sees it been 'an historic receptacle' for tyrannies, and as a consequence 'a ground always open to new masters'.[6] The new masters had on this occasion espoused a creed which served as a powerful justification for domination, not only over 'the Eurasiatic steppes' themselves, but also over the whole of Europe. De Chessin's concept of 'Mongolian socialism' was akin to Kristof's explanation of Russian behaviour in Asiatic terms. 'They are the non-nomadic heirs to the nomads; they have rebuilt the Mongol empire from its western end.'[7]

While subscribing to the idea of the uniqueness of the Soviet Union, Timasheff unscrambled its components and detached its central feature from the Russian past. 'The formation of the Soviet Union', he writes,' was entirely at variance with the historical pattern of Russian expansion'.[8] As a result of the unique nature of the new ideology, 'its limit is the totality of the globe', something which Timasheff contends was never in the thoughts of 'historical Russia'.[9] Prior to the Revolution he sees the objective data as confirming that 'Russia was no more aggressive than any other great power'.[10]

However, whether they derived from broad geopolitical perceptions or narrower ideological ones, the very real fears expressed by Lavisse of 'the Russian glacier... always gliding onward'[11] inevitably came to influence international views and to bring about defensive reactions. 'Soviet Russia' became a composite term for something which was both a tangible reality and the expression of a disquieting ideal, the fused elements of which included its size, location, resources, nationality, ideology and system of government. Its very immensity, as Mahan had said of the Russian Empire, was calculated to arouse feelings of awe and apprehension. While this apprehension had been present well before World War II, the spectacular Soviet victory over Nazi Germany reinforced it. To Weigert the Soviet Union was a Eurasian state with almost unlimited potential power, and he quoted the prophetic Spengler who 'saw clearly the gigantic contours of the Asiatic face of Russia arise in the mists that cover the future'.[12] The 'aziatchina', which to Bukharin was the 'real living spirit of the revolution' representing true liberation from Europe, was from the western geopolitical viewpoint a hostile and menacing spectre rising out of the east.[13]

In the years following World War II western fears aroused by the Soviet Union's 'gigantic contours' were clarified and rationalised by Anglo-American political geographers. In doing this a large number of them looked at Mackinder's Heartland theory. Weigert affirmed in 'Heartland Revisited' that in his estimation, 'Mackinder's citadel of land power still stands - and mightier than ever.'[14] He was the first in a long line of post-war political geographers who then paid their respects to the Heartland theory and vouched for its continuing validity in the contemporary world. This post-war Heartland would not in every case have been recognisable to Mackinder himself, since its size and shape was often much altered by those who then addressed themselves to it.[15] It was greatly enlarged by Fawcett to cover areas of Africa and the Middle East, and was shrunk by Meinig into the fast-nesses of central Asia, but either way it retained its position as the ultimate focus of land power and as such an enduring danger to the maritime world. Gilbert and Parker were in no doubt as to its 'continuing validity'[16] and even the more sceptical East and Spate, while questioning many of Mackinder's assumptions, were conscious of 'its ominous ring and uncanny relevance to the present world situation'.[17] Walters, however, remained unconvinced and saw the Heart-land as being less a geopolitical reality than a 'dreadfully ominous' symbol. In endeavouring to explain this symbolism he invoked 'The Castle' by Franz Kafka. 'The silent castle', he said, 'epitomised by the Kremlin itself and by the Heartland, loomed overpoweringly in the dark recesses of the mind. It is no wonder that the writings of Kafka seem to reflect the atmosphere of the Cold War.'[18] By the 1950s so strong had become the strategic fixation with the Heartland that Stephen Jones commented that it appeared to have been transformed into a fatalistic doctrine of 'Herzland über alles'.[19] The underlying reasons for this were to be found less in its real attributes than in its role in the international scene as per-ceived from the West since World War II. This was dominated by the global confrontation of the two antagonistic super-powers, and the Heartland theory took on a new lease of life in this context. As a result it was seized upon as a method of giving a new spatial explanation for the world scene.[20]

As has been observed, the Soviet Union exhibits a close conformity with many of the characteristics of the dominant state as specified in Chapter 3. In the evolution of its core region, directions of territorial expansion, internal movements of power, acquisition of 'limites naturelles' and expansion towards the sea its spatial behaviour has been akin to that characterising the other dominant states which have been examined.[21] Nevertheless, there do remain a number of geopolitical characteristics which have contributed to setting the Soviet Union apart in the minds of foreign observers. The most significant of these concern its size, location, territory and concept of territorial authority.

The sheer dimensions of the country defy the imagin-ation. It still appears to have that 'limitlessness' and 'boundlessness' on which so many writers have commented. Its total area of 22 million square kilometres makes it some 45 times the size of France and 93 times that of Great Britain. it remains the 'vast uninterrupted mass' which Mahan observed

Table 3:   Comparative Figures for Area, Population and
           Primary Production for the Soviet Union, France,
           West Germany and the United Kingdom

	Soviet Union	France	West Germany	United Kingdom
Total area (Mill.sq.km)	22.402	0.544	0.249	0.244
Population (mills)	275	54.9	61.2	56.5
Total annual primary energy output (Mill. tonnes oil equivalent)	1493.9	74.0	124.0	203.9
Iron ore (Fe content mill. tonnes)	147.1	5.2	0.3	0.1
Total wood output (Roundwood 1000m³)	353,900	28,342	29,000	4,088
Total wooded area (1000 hectares)	932,000	14,618	7,360	2,273
Total land area in agricultural use 1000 hectares)	605,415	31,445	12,019	18,644

Source: Eurostat, (Statistical Office of the European
          Communities, 1986)

at the beginning of the twentieth century, with seemingly no
obstacles 'to impede the concentrated action of the disposable
strength'.[22] This strength is also considerable. The Soviet
Union's total population of 275 millions is five times that of
either Great Britain or France and over three times that of
both German states combined. At the time of its greatest
territorial expansion in the middle of the sixteenth century
the Ottoman Empire had an area of only one-third and a
population of one-tenth that of the Soviet Union today. The
resources and output of the country are also enormous

compared to those of the other powers in the western ecumene (Table 3). Its total annual output of energy is around 1500 million tonnes oil equivalent, which is twenty times that of France, twelve times that of West Germany and over seven times that of Britain which is the largest producer of energy in the Maritime Crescent. Its total forested area is over 900 million hectares which is 65 times that of France and 450 times that of the United Kingdom. While west of the Urals the country has been intensively exploited, east of the Urals huge reserves of energy, metals and wood remain. Siberia, said Kirby is a huge '"lucky dip" for resources containing enormous prizes'.[23] In terms of area, population, primary production and resources, the Soviet Union is thus of a different order from the other states of the western ecumene.

Its size, together with its central location, has also been considered as giving the Soviet Union a strategic uniqueness. Cressey saw this great size as having 'made it possible for her to sell space with which to gain time'.[24] Weigert for his part asserted that 'Russia is unconquerable from outside' because 'Distance is a force which has not yet been conquered'.[25] Such views arose in part out of Heartland thinking and the association of the Soviet Union's strength with it. Images of the 'citadel', the 'interior fortress' and the 'geographically charmed sanctuary' indicate a strong sense of the invulnerability of the Soviet Union and its position as a sort of ultimate world fortress, a 'Festung Herzland'. Of course, a fortress may well be invulnerable, but this same invulnerability could also make it a safe base from which to attack neighbouring states. In this way the Heartland appears to bestow a great strategic advantage on its tenant. In Neville Brown's words, 'the nodality of the U.S.S.R.'s north Eurasian location gives it a potential for military leverage virtually all round a circumferential zone from Finnmark through Iran and Afghanistan to Manchuria'.[26] In this Eurasian context, the European Maritime Crescent takes on the appearance of being both territorially insignificant and strategically weak, all too vulnerable to the unwelcome attentions of its formidable neighbourhood superpower.

The third characteristic is the particular territorial qualities of the Soviet Union. This centres on de Chessin's 'immense shapeless plain stretching between two worlds'. The country's history has been played out in this natural isolated amphitheatre bounded by mountains, deserts, swamplands and ice-covered seas. Physical isolation has helped to create that 'psychological distance' which had the effect of mutating and changing those ideas which seeped in. Ideas originating in the west were, as Liska saw it, 'intensified ... or deformed in the East', and it was the 'mutation' arising from this which then produced eastern Marxism.[27] Great cultural changes in the west, such as the Renaissance and the Reformation, largely passed Russia by. In her vastness, observed Peter

THE SILENT CASTLE

Figure 7.1: The Soviet Union and the Western Ecumene

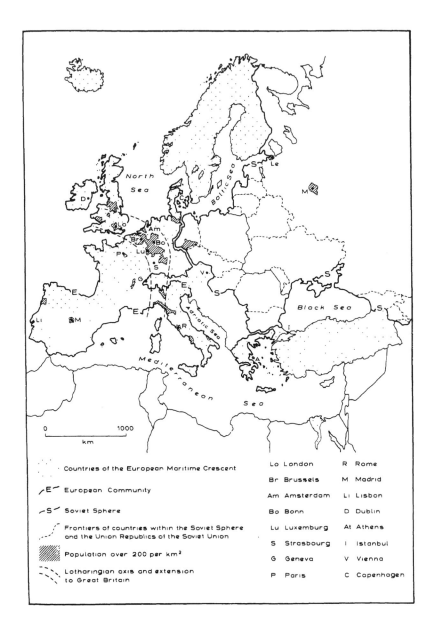

Ustinov, 'she heard no voices from the next room'.[28] In this remote, isolated and relatively homogeneous environment a strong cultural individuality emerged. The attempt to express this individuality and to develop on it was an important part of the Soviet Union's aziatchina. Bestriding the continents, it came to be seen as being what Kristof termed 'an "organically" united world, pulsating with a civilisation, superior and above all "our own"'.[29] Both inside and outside the Soviet Union this civilisation was seen as a challenge and, ultimately, an alternative to the west.

Finally there is the enduring quality of the state which rules this 'middle world' and which traces its lineage back through Kievan Russ to Byzantium and Rome. As has been seen, claims to an imperial lineage which legitimate the position of dominance have characteristed all those states which have made bids for supremacy over the western ecumene. However, the state which over the centuries has dominated the great plain from the fluctuating borders of Christendom eastwards to the Urals has certainly been outstanding in its durability. The existence of a single state, an imperium, unifying this immense area has become almost as much a part of its territorial reality as are the rivers which flow over it. The plain may have in one view been 'an historic receptacle' of tyrannies, but these tyrannies have also imposed upon it a structure of order and authority based on a world-view onto what otherwise could well have degenerated into fluidity, anarchy and chaos. Mahan's 'vast uninterrupted mass' had proved inimical to stable political units of more modest size such as those characteristic of Henrikson's 'neatly segmented, multicoloured world'.[30] To Whittlesey its very size 'made Russia the natural protagonist of the centripetal point of view' and he saw this as being opposed to 'the centrifugal view of maritime western Europe.'[31] Until 1917 this centripetal view had been a theocratic one based on Orthodoxy, but the Revolution had the effect of transforming it into a materialist one based on Marxism. There was no change, however, in the belief that ultimate control should be vested in a central authority. Whether this was symbolised by the Basileus, Samoderzhets, Tsar or Vozhd, the unbroken lineage of central authority had become virtually a part of the natural order of things.[32] In the Soviet interpretation of Russian history, periodic disunity is seen as having been one of its most calamitous features. Such periods resulted in the widespread breakdown of law and order and, even more injurious, the incursions of outsiders from both east and west into what they saw as the all too vulnerable Russian lands.

These four geopolitical characteristics prima facie support the contention that the Soviet Union possesses both the means and the motives for further expansion and ultimately the accomplishment of what some observers, such as Pap, have been convinced was 'the pole star of its policy', namely 'world domination'.[33] However, while in the context of the western ecumene alone the Soviet Union's uniqueness is difficult to dispute, in a world context it is far more open to question. The Soviet Union is certainly the largest state in the world, but its lead over other powers of continental dimensions is not anything like as great as that over the European powers. Thus while it is 45 times the size of France, it is only 2.6 times the size of Brazil and 2.4 times the size of the United States. Its total population is actually considerably lower than those of either China or India and only slightly larger than that of the United States. In respect of energy production, the Soviet output of coal is below that of the United States, its hydro-electricity production is less than either the United States or Canada and its output of oil is below those of the countries of the Middle East. As to agriculture, both the extent of its agricultural land area and its output is mediocre compared to many other countries of smaller size. While it undoubtedly possesses enormous un-tapped resources in Siberia, there is little evidence that these are of a completely different order from those to be found elsewhere. The fact is that in size, population and physical resources the Soviet Union compares variably with the other major centres of power, both actual and potential. There appears to be little real evidence for the existence of some uniquely abundant internal El Dorado.

Then there is the idea that the Heartland is a sanctuary of land power, which provides a base for attack but which cannot itself be successfully assaulted. While the Heartland was certainly invulnerable to sea power, air power began to develop the capacity to erode this invulnerability at the time of World War II and to penetrate into its fastnesses. Since that time the enhancement of air and missile power has ended this invulnerability for good. Since the 1960s no part of the Soviet Union has been out of the range of either the seaborne or land missiles deployed around its peripheries. Its sheer size does make individual targets within it more difficult to pinpoint, but satellite surveillance of Soviet territory has now made even this less of a problem. The Heartland, until recent times so mysterious and unknown, has been able to retain few of its secrets. As to the strategic advantage of that 'nodal' location noted by Brown, these are bound to be relative and to depend on the state of the balance of power at any time. What appears to be 'nodality' from one viewpoint can mean 'encirclement' from another. What from the standpoint of a western strategist is the capacity of the Soviet Union to strike over a 'circumferential zone' may from a Soviet stand-point be vulnerability to attack over the same wide area. The location of the Soviet Union in relation to the World-Island bears many resemblances to that of Germany in relation to the western ecumene. The German answer to the 'curse' of its central location was to increase the strength of Mitteleuropa

THE SILENT CASTLE

so as to make it more powerful than its peripheral opponents and thus gain security through strength. This was a strategy which ultimately failed because the physical power available to the maritime and continental peripheries was ultimately greater than that available to land-locked Mitteleuropa. By analogy with this early twentieth century European scenario, 'Festung Hertzland' could well be no more secure against the united forces of the peripheral powers than had been 'Festung Europa' at the end of World War II.

The uniqueness of the Russian 'middle world' bestriding Europe and Asia is also open to question. Russia occupied and colonised large parts of northern Asia as an imperial power, and to her it was just as much colonial territory as that of other European powers at the time. The nature of the Byzantine inheritance, the oriental half of the Roman Empire, and the process which Kristof described as the 'recreation of the Mongol Empire from its western end' were unique phenomena only in the context of the western ecumene. Intercontinental contacts and those cultural symbioses which are the result of them, have become far more widespread during modern times. The concept of the two unbridgeable worlds, western and eastern, has become far less convincing in the age of globalism. Such a stark dichotomy, which seemed so real even to Mackinder, is replaced by a diversity of culture areas, each one having its own particular set of characteristics arising from the fusion of its indigenous culture with imported traits resulting usually from commercial and political contacts. In Asia itself, both India and Japan are examples of countries which have experienced the grafting of western shoots onto the trunk of their indigenous civilisations. As a result something different, and in its own way 'unique' has been created.

Finally there is the contention that this state is unique by virtue of its durability and the way it has brought unity and order to an enormous geographical territory for such a long period. Again however, while in the context of Europe, such durability is indeed remarkable, in the context of Asia it is much less so. In the eastern and southern ecumenes, the Far East and the Indian sub-continent, the existence of extensive states has been a normal phenomenon of the political map. These have usually dominated most of the ecumene itself, and incorporated large additional territories deep into the heart of Asia. In both China and India 'Generalls of the World', whether of Manchu or Moghul origins, have successfully established vast empires, subsequent versions of which have through the ages continued to control the same territories.

Thus on the basis of the evidence yielded by a geopolitical examination, the conclusion emerges that, as with the rumours of the death of Mark Twain, those of the uniqueness of the Soviet Union have been greatly exaggerated. What from a western perspective may have the appearance of being a uniquely large and dangerous phenomenon, from a Eurasian or Asiatic perspective appears to be far less so. As Liska comments, 'The underlying secular tendency is for the more west-lying power to view the day's main east-lying power as one that draws on a more-than-actually powerful base for hatching mysterious schemes in support of more-than-life-size expansion'. From this he concludes that 'Western perceptions grew more suspicious and the threats more ominous as the target power moved eastwards'.[34] Liska's observations certainly accord with that 'Tamburlaine syndrome' in which the blood-soaked eastern conqueror strikes westwards across the steppes in pursuit of a dream of world domination. Nevertheless, side by side with this persistent phobia, the states of the European Maritime Crescent have also felt more specific fears of that power which has been able to dominate, or threaten to dominate, the centre of the continent. During her period of dominance Spain, lying to the west, was more feared than France at the centre. Subsequently France became more feared than Prussia, lying further to the east, and the German Empire was more feared than Russia lying yet further to the east of her. Real fear has been engendered in these cases far less by the mystique with which a particular location has been endowed, than by the perception of the rise of a power which possesses both the will to dominate and the resources with which to make it possible.

Since World War II the Soviet Union has appeared to possess them both. The voluntary union of socialist republics became transformed into a centralised state and the threatened bastion of socialism was converted by victory into a formidable world power. This was accompanied by a mental retreat from the world into a kind of Communist Festung characterised by xenophobic aggressiveness of which Stalin, the 'Vozhd', was in many ways the symbol. The Soviet Union turned away from what it perceived as being a hostile and ungrateful Europe to create 'socialism in one country' in the new lands of Eurasia. By analogy with the proclamation of the Italian Risorgimento that 'l'Italia fara de se', Kirby concludes that 'Russia will "do its own thing", and Siberia is very much the big and fresh place in which to do it'.[35] This powerful assertion of national self-confidence coupled with defiance of the west, is encapsulated in Alexander Blok's poem 'The Scythians':

> You have your millions, we are numberless,
> numberless, numberless. Try doing battle
> with us! Yes we are Scythians! Yes,
> Asiatics, with greedy eyes slanting.[36]

This 'skifstvo' (scythianism) is a version of 'aziatchina' and both concepts express the idea of rebirth and renewal taking place on a massive scale in Eurasia.

Yet, in the end one must come to the conclusion that it has all been heard before. Attitudes of this sort are traditionally part of the baggage of those states which in the past have made their bids for positions of dominance. Beneath the gigantism, the xenophobia, the 'scythianism', the ideology and the apocalyptic assertions of the coming of a new age for mankind there lies little substantial geopolitical uniqueness. The Soviets may be Eurasians rather than Europeans, but they are most certainly not 'numberless', nor are they in other respects fundamentally different from other states. In the context of the western ecumene, the Soviet Union's residual uniqueness is in its being its principal 'east-lying power' and as such constituting a physical link between Europe and Asia, between the western ecumene and the eastern one.

It now remains to be considered whether further dominating behaviour is therefore to be anticipated from the Soviet Union, along the lines suggested by the model, or whether certain internal changes now taking place in the country's geopolitical structure may give indications of a further change in its international behaviour.

NOTES AND REFERENCES

1. R. Niebuhr, *Nations and Empires. Recurring Patterns in the Political Order* (Faber, London, 1959), p. 217.

2. S. de Chessin, *Darkness from the East* (Harrap, London, 1930).

3. S. de Chessin, ibid., p. 193.

4. S. de Chessin, ibid., p. 8.

5. S. de Chessin, ibid., p. 251.

6. S. de Chessin, ibid., p. 247.

7. L.K.D. Kristof, 'The Russian Image of Russia: An Applied Study in Geopolitical Methodology' in C.A. Fisher (ed.), *Essays in Political Geography* (Methuen, London, 1968), p. 363.

8. N.S. Timasheff, 'Russian Imperialism or Communist Aggression' in W. Gurian (ed.), *Soviet Imperialism. Its Origins and Tactics* (University of Notre Dame Press, Notre dame, Indiana, 1953), p. 34.

9. On the contrary it had been very much in the thoughts of historical Russia. From the 'Third Rome' of the sixteenth century to the Pan-Slav movement of the nineteenth there had been many who saw, in the poet Tyutchev's words, 'The world-wide destiny of Russia'. See W. Kolarz, *Myths and Realities in Eastern Europe* (Drummond, London, 1946), p. 84.

10. This is also confirmed by the 'objective data' exam-

ined in Chapter 6. Where Timasheff goes wrong is in ascribing uniqueness to Russian messianism. Messianism of one sort or another has in fact been a characteristic of almost all the dominant states of modern times.

11. E. Lavisse, *A General View of the Political History of Europe* (Longmans Green, London, 1891), p. 159.

12. H.W. Weigert, *Generals and Geographers. The Twilight of Geopolitics* (Oxford University Press, New York, 1942), p. 46.

13. 'The *aziatchina* was the real living spirit of the Revolutions; it was Marxism which does not fall on its knees before the state power of capitalism'. Quoted in L.K.D. Kristof, 'The Russian Image of Russia', p. 378.

14. H.W. Weigert, 'Heartland Revisited' in H.W. Weigert et al., (eds.), *New Compass of the World: A Symposium on Political Geography* (Macmillan, New York, 1949), p. 89.

Later on, however, Weigert did give the warning that 'we must be careful not to be caught napping in the nineteenth century', See H.W. Weigert et al., *Principles of Political Geography* (Appleton-Century-Crofts, New York, 1957), p. 211.

15. In fact, by the time of the renewed interest in the subject, Mackinder himself had altered his views to such an extent that R.E. Walters was of the opinion that he had, in effect, abandoned the Heartland altogether. See H.J. Mackinder, 'The Round World and the Winning of the Peace', *Foreign Affairs*, XXI, no.4, 1943. In this article Mackinder did still contend, however, that 'the Heartland is the greatest natural fortress on earth'.

16. E.W. Gilbert and W.H. Parker, 'Mackinder's "Democratic Ideals and Reality" after Fifty Years', *Geographical Journal*, CXXXV, no.2, 1969, pp. 228-31.

17. W.G. East and O.H.K. Spate, *The Changing Map of Asia* (Harrap, London, 1961), p. 357.

18. R.E. Walters, *The Nuclear Trap: An Escape Route* (Penguin, Harmondsworth, 1974), p. 43.

19. S.B. Jones, 'Global Strategic Views', *The Geographical Review*, XLV, no. 4, 1955, p. 496.

20. G. Parker, *Western Geopolitical Thought in the Twentieth Century* (Croom Helm, London, 1985), p. 133.

21. The urge to the sea has been one of the most discussed of all the geopolitical explanations of Russian and Soviet spatial behaviour. As W.H. Parker put it, 'A thousand years of Russian history can be interpreted as that of a land-locked state struggling to break out from claustrophobic isolation'. See W.H. Parker, *Mackinder. Geography as Aid to Statecraft* (Clarendon Press, Oxford, 1982). In many ways *Wasser über alles* has been ahead of *Herzland über alles* as an explanatory factor. It is perhaps a reflection of the search for the one definitive explanation of Soviet conduct which will illuminate all the rest. While it has certainly been a powerful

motive, it is certainly not unique to either Russia or the Soviet Union. As has been observed, it was an important element in the behaviour of both the Ottoman Empire and Germany and was also present in the other dominating states. Neville Brown implicitly recognises this in his discussion of 'the warm water ports thesis' by asking, 'Was there anything in the time-honoured contention that, rather like Berlin under Kaiser Wilhelm, Moscow seeks to escape from the confines that physical geography imposes upon her?' See N. Brown, *Limited World War* (Canberra Papers on Strategy and Defense, no. 32, Australian National University, Canberra, 1984), p. 23. 'Escaping from the confines' is, of course, an essential element in the expansionist process. As has been seen, the principal thrust of Russia was southwards towards the cultural macrocore with all that this implied. See also Chapter 5 and W. Kolarz, *Myths and Realities in Eastern Europe*, pp. 83-5.

22. A.T. Mahan, *The Problem of Asia* (Little, Brown & Co., Boston, 1900), p. 24.

23. S. Kirby, 'Siberia: Heartland and Framework' in C. Keeble (ed.), *The Soviet State. The Domestic Roots of Soviet Foreign Policy* (Gower for the Royal Institute of International Affairs, London, 1985), p. 153.

24. G.B. Cressey, 'Siberia's Role in Soviet Strategy' in H.W. Weigert and V. Stefansson (eds.), *Compass of the World. A Symposium of Political Geography* (Harrap, London, 1943), p. 23.

25. H.W. Weigert, 'Heartland Revisited' in H.W. Weigert et al., *New Compass of the World*, p. 80.

26. N.Brown, *Limited World War* (Canberra Papers on Strategy and Defense, Australian National University, Canberra, 1984), p.14.

27. G. Liska, *Russia and World Order* (Johns Hopkins Press, Baltimore, 1980), p. 63.

28. P. Ustinov, *My Russia* (Macmillan, London, 1985), p. 197.

29. L.K.D. Kristof, 'The Russian Image of Russia,' p. 384.

30. A.K. Henrikson, 'The Geographical "Mental Maps" of American Foreign Policy Makers', *International Political Science Review*, vol. 1, no.4, 1980.

31. D. Whittlesey, *The Earth and the State. A Study of Political Geography* (Holt, New York, 1939), p. 535.

32. '*Samoderzhets*' is a Mongol word meaning 'ruler' or 'wielder of supreme authority'. It was subsequently adopted by the Princes of Muscovy to indicate that they had inherited the authority of the Mongols in the Russian lands. Its continued use was a stiffening to the theocratic state structures derived from Byzantium. '*Vozhd*' is the Russian word for leader and, at a time when the word 'Leader' in many languages was being used to denote the heads of authoritarian

states elsewhere in Europe, it was adopted by Stalin. His successors tended to revert to more innocuous terms such as 'General Secretary' and 'Chairman' to soften the impact of centralised authority.

33. M. Pap, 'The Ukrainian Problem' in W. Gurian (ed.), *Soviet Imperialism. Its Origins and Tactics* (University of Notre Dame Press, Notre Dame, Indiana, 1953), p. 54.

34. G. Liska, *Russia and World Order*, p. 104.

35. S. Kirby, 'Siberia: Heartland and Framework', p. 151.

36. Alexander Blok, 'The Scythians' in *The Twelve and Other Poems*, trans. Jon Stallworthy and Peter France (Eyre and Spottiswoode, London, 1970), p. 161.

EIGHT

DECLINE AND FALL: THE RETREAT FROM DOMINANCE

To Robert Walters 'The Silent Castle', as epitomised by the Kremlin itself, loomed overpoweringly in the dark recesses of the mind, an image of awe and dread to all who lived within its shadow.[1] Kafka's Castle, from which Walters' evocative concept was derived, showed not the slightest sign of life when viewed from a distance and, even as K looked at it, '(its) contours were already beginning to dissolve'.[2] From the geopolitical viewpoint, the 'Castle' is the crystallisation of those fears which have been engendered by dominant states throughout history. However menacing and enduring they may have been at the time, they too have eventually 'dissolved' to be replaced by new ones. According to Goblet the process of decline and dissolution was not separate, like a disease affecting the body politic, but was inherent in the nature of the dominant or 'extensive' state. He held the opinion that 'the dissolution of the extensive state begins as soon as the initial period of conquest and occupation is over'. The reason for this, as he sees it, is that 'Territorial expansion is vital to their very existence and they can think in no other terms but of an extensive exploitation.' When this expansion is brought to a halt, 'so ends the period of conquests, so begins the period of disintegration'.[3]

The actual scenario for the end of the period of dominance of a particular state has involved the build up of countervailing power, usually in the form of a coalition of states, which has then successfully challenged and defeated the dominant state of the day. However, the failure of the dominant state to respond successfully to this external challenge, after in the past having won so many victories over just such opposing alliances, needs some further explanation. Historical explanations have tended to emphasise the deterioration in the quality of the leadership of the dominant state. According to this view, decline takes place because lesser men are now at the helm than during that 'golden age' of power when her triumphant armies swept on from one victory to another. Such men, often born to ease and privilege, are far less capable of

147

steering the ship of state successfully than were their illustrious predecessors.

The question of the actual role of the leadership, and in particular that of the charismatic leader, is a complex one. Much of traditional history has been dominated by such leaders, and their exploits have given it much of its excitement and its 'story' quality. The heroic exploits of Napoleon are bound to be far more exciting than is that rapid growth of the French population which made his massive armies possible. Likewise the deeds of Hitler and the leadership of the Third Reich arouse more fascination and dread than the coal and iron production figures of the Ruhr which sustained the Nazi bid for dominance. Yet geography is bound to be the real foundation of national power and, as Bismarck observed, it is 'blood and iron' which secure its interests. As has been seen, the nature and raison d'être of the typical dominant state is forged in the harsh environments of remote core regions long before the advent of the charismatic leader who is to take it on from victory to victory. It is the core and not the leader which determines the character and the direction which the eventual bid for dominance will take. As Tolstoy put it, Napoleon did not lead his army into Russia, but rather he followed it. From the geopolitical viewpoint, the leader is in many ways a result and not a cause, and the 'golden age' is rooted in a favourable set of geopolitical circumstances. As with its rise, the decline of the dominant state is characterised by profound changes in its geopolitical structure. Whatever the qualities of the leadership may be, in the last analysis it is these changes which make it impossible for the state any longer to maintain its dominant position. Those geopolitical characteristics associated with the decline of the previous dominant states have been itemised in Table 4, and their individual and overall incidence is specified. When such characteristics are to be found in the Soviet Union, these too have been noted in the Table.

The process of expansion has eventually taken the dominant state into unfamiliar and hostile physical environments. These combine with distance from the principal centre of power to make the grip of the dominant state weaker. Thus the Ottomans and the Austrians in the Balkan mountains, the Spaniards in the tempestuous seas of northern Europe and the French and the Germans in the icy wastes of Russia were confronted with combinations of physical conditions with which they were totally unable to cope. The peoples native to these environments, and who had evolved their particular 'genres de vie' within them, could not be permanently held down by intruders from other physical environments who were ill-equipped to cope with the local conditions. While the regions close to the principal centres of power had been subdued with a minimum of cost and effort, the physical and human resources required to control the more remote and difficult areas became progressively greater than those available to the dominant state. The end result of persistent refusal by successive dominant states to accept this was catastrophic and costly defeats such as that of the Spanish Armada in the North Sea, the grande armée in the retreat from Moscow and the Wehrmacht in front of Stalingrad.

The response of the rulers of the dominant state to the problems of maintaining its position of ascendancy was to increase the centralisation of control. Thus within the vast Spanish Empire control was centralised first on Madrid and then on El Escorial. Likewise in France the comprehensive and sophisticated structure of political control centring on Paris was modernised and then extended to the conquered territories. The Austrian Empire and the Third Reich, both heirs to the centrifugal structures of the Holy Roman Empire, also attempted, though with far less success, to impose centralised authority on their vast dominions. A yet more basic method of attempting to secure control was the attempt to diminish diversity and thus make the population more homogeneous. This was associated with the persecution and expulsion of those elements within the population which were considered to be subversive and dangerous to the cohesion of the state. In this way over the centuries and in various states large numbers of Moors, Jews, Protestants, aristocrats, Armenians, Communists and Slavs have, in one way or another, been removed in the alleged interests of the security of the state and the purity of its ideals. Since it was invariably physically impossible for the dominant state to eradicate all dissent, the result of such actions was inevitably the opposite of what had been intended. It was to increase rather than diminish the dissatisfaction of large sections of the population. Such dissatisfaction became widespread and tended to be expressed in political, social or economic ways. It was generally strongest in the peripheral regions where it usually crystallised into some form of regional or national movement. In this way nationalism developed into a powerful alternative ideology to that propounded by the dominant state. What made this possible was what Goblet called 'the persistent survival within its territory of the intensive cores of old states', which were then reborn out of conditions of adversity.[4] It was expressed in the refusal of the peripheral lands of the Iberian peninsula to accept total submission to their Castilian masters; in the refusal of the Poles to accept assimilation within the bloated empires of Prussia, Austria and Russia; in the refusal of the Hungarians to accept a role subordinate to the Austrians within the multi-national Habsburg Empire.

A further characteristic of the period of decline concerns the spatial distribution of economic power. This entails a shift of the state's economic centre of gravity away from its historic core to a new economic centre located elsewhere in its territory. Such a shift is the result of general technical

Table 4: The Retreat from Dominance

Characteristics	Ottoman Empire	Spain	Austria	France	Germany	Soviet Union
1. Territorial expansion into hostile environments	x	x	x	x	x	
2. Expansion halted in these environments	x	x	x	x	x	
3. Increased attempts at policies of centralisation	x	x		x	x	x
4. Attempt to impose overall geopolitical structure		x	x	x	x	x
5. Regional and national hostility to dominant state	x	x	x		x	x
6. Shift of economic power centre away from historic core	x	x	x	x	x	x
7. Move of power (economic, demographic and cultural) towards subject nationalities	x	x	x			x
8. Move of power to peripheries		x	x	x		x
9. Transfrontier contacts by populations of peripheries	x	x	x		x	x
10. Attempt to introduce greater homogeneity in population	x	x		x	x	. x
11. Resurgence of core nation nationalism	x		x	x	x	x
12. Final bid for supremacy fails	x	x	x	x	x	
13. Failure to subdue native peoples	x	x	x	x	x	x
14. Distintegration of dominant state into components	x	x	x	x	x	
15. Position of dominance comes to an end after war	x	x	x	x	x	
Percentage incidence of characteristics	86	93	86	80	86	83*

* percentage of characteristics other than those pertaining to actual end of position of dominance

advances such as the development of new sources of energy and raw materials and of improved methods of transport. As a result of this an entirely new centre of population emerges which is likely to have very different social and cultural values from those of the core state. A wider political gap may also open up between this new commercial-industrial centre and the core state since the former is likely to be unsympathetic to the latter's aspirations and may well see its own interests in quite a different light. The power of dynamic regions such as the lower Guadalquivir, the principal focus of Spain's maritime empire, and the Rhine-Ruhr conurbation, 'the anvil of the Reich', have had the effect of weakening the grip of their respective core states. In the Ottoman Empire the Marmara region early on took over the position of core region from Turkish Anatolia. The difference here was that the whole of the apparatus of the state was moved to the new region and a process of cultural fusion then began. This early fusion of the political, economic, demographic and social cores was a feature unique to the Ottoman Empire and became a cause of that lack of dynamism or innovation as a result of which it fell further and further behind the western states. It was a victim of that 'stagnation, epigonism and retrogression' which Wittfogel identified as being a particular feature of oriental imperial states in decline.[5] While in the dominant states further west the continued geographical separation of the political from the economic centres of power was associated with the continuance of economic dynamism for a longer period, they too later moved into a condition of entropy relative to other smaller states. This came about particularly as a result of innovation taking place beyond their frontiers which had the effect of relegating them to a position of economic inferiority. Thus, in the economic sphere, Spain gave place to northern Europe in the later sixteenth century, France gave place to Britain in the later eighteenth century, and Austria and France both gave place to the new German Empire in the nineteenth century. Those internal El Dorados upon which the growth of power had originally been based thus proved inadequate to sustain it in the face of the emergence of greater power elsewhere.

A notable feature of internal spatial change associated with decline is a shift of power from the centre towards the peripheries of the state. This arises from changes in the internal economic structure, but it eventually leads on to wider demographic, social and financial changes in favour of the peripheries, and these in turn have considerable repercussions on the balance of political power within the state. Thus, during the period of the decline of Spain, there was a marked shift in power from the Meseta to the Iberian coastlands; in France from the Paris basin eastwards towards the Rhine and in Austria from the middle Danube to the sub-Carpathian belt. This had the concomitant effect of increasing the wider importance of the nationalities living in these areas and in this way Catalans, Andalusians, Czechs and the inhabitants of Alsace-Lorraine came to play a more significant role in the affairs of their respective imperial states. With the increase in prosperity came the opening up of trans-frontier economic contacts, and this in turn facilitated the development of trans-frontier cultural and political contacts. As a result, the South Slavs of the Balkan mountains began to distance themselves from their various imperial masters and to look towards a Pan-Slav future in which their long-suppressed national identities would emerge and flourish once more. Long gone frontiers were resurrected in the minds of nationalists in preparation for the coming of the day when they could once more be converted into realities.

The other characteristics of decline are related to the failure of the final bid by the dominant state to hold on to its vanishing position of supremacy. The formidable combination of the build-up of opposing power together with growing internal weakness and division becomes too much for it. As the outlying parts of the empire then begin to break away, a resurgence of nationalism takes place within the core nation itself. There is a nostalgic turning away from the declining empire towards the nation itself and its heroic and semi-mythical past. In both the Ottoman and the Austro-Hungarian Empires strong Turkish and German-Austrian nationalism emerged during the period of advanced decline. Likewise modern French nationalism was an outcome of the Revolution which terminated her period of dominance, although France still remained sufficiently strong to retain a quasi-imperial grip on a number of small subject peoples, in particular the Bretons, Provencals and Alsace-Lorraine Germans.

Thus clear indications of fundamental geopolitical change can be identified as characterising the period of decline of the dominant state. The average overall incidence of the fifteen geopolitical characteristics itemised in Table 4 is 87 per cent. The actual individual incidences range from France with 80 per cent to Spain with 93 per cent, with Germany, Austria and the Ottoman Empire each having incidences of 86 per cent. The most important overall effect of these changes is to undermine the degree of unity which was created during the period of ascendancy. They exacerbate the central geopolitical problem of dominance and the ultimate source of its weakness - the seemingly endless appetite for territory. 'It attempts to dominate territories', comments Goblet, 'of a size disproportionate to the power which it is able to wield in their government', and as a result of this 'it lives permanently in a state of unstable equilibrium, whose upsetting leads to ... the frequently abrupt dissolution and destruction of the state'.[6] The extent to which such characteristics of decline can also be identified in the Soviet Union will now be considered.

In respect of the first two characteristics in Table 4, there is no conformity in the case of the Soviet Union. Neither Soviet nor Russian expansion before it was brought to a halt by a hostile physical environment. The physical conditions of the Soviet core regions are themselves among the most harsh in the whole of the western ecumene, and expansion actually took the state into more congenial conditions.[7] Nor was territorial expansion brought to a halt catastrophically as had been the French and German expansion before it. Rather, as Lavisse put it, the glacier 'glided onward', inexorably adding yet more territory. After being pushed back to some extent after World War I, it proceeded to glide forward again after World War II. However, as with the other dominating states under discussion, the smaller nationalities which lay in its path were, for the most part, hostile to it. This hostility was considerable on the part of those nations which now found themselves back in the fold after an all too brief period of independence, and there were also manifestations of it with the Soviet sphere in eastern Europe. Despite an awareness of such reactions, and the ostensible confederal structure of the Soviet state, which in theory allowed for such diversity, the actual policy remained one of the maintenance of rigid centralisation on Moscow. The basic aim remained the creation of an ideological rather than a national identity, but the two things were difficult to separate in practice. In the alleged interests of the security of the Soviet state and the purity of its ideals, large numbers of class enemies were purged, and in the wake of World War II nationalities alleged to have been less than loyal were harshly dealt with, many of them being geographically displaced away from the strategically vulnerable peripheral regions into remoter parts of the country. This was an implicit recognition of the possible dangers arising from a cultural dissociation of the peripheries from the core nation. In place of the proliferation of national groupings, each having its own local loyalty, came the attempt to create 'Soviet man' out of a single class, a single belief and one dominant culture.

The movement of industrial strength out from the Russian core and southwards to the Ukraine had preceded the Revolution, but the relative industrial strength of the Ukraine was further increased by Soviet planning policy and the heavy industrial base was then bifurcated with the replication of the Donbas-Krivoi Rog region in the Urals-Kuzbas complex. These developments were followed by the rise of power in the peripheries, and their growing contribution to the whole Soviet sphere. In particular the countries of eastern Europe north of the Carpathian divide, the Baltic republics and the regions around the Black Sea rapidly gained in relative economic and demographic strength. These same peripheries also began to take an increasing interest in the regions lying beyond the frontiers of the Soviet sphere and this interest often had a cultural as well as an economic character. Thus the German Democratic Republic has looked towards the Federal German Republic, Hungary has looked towards her old imperial partner Austria, Estonia has looked towards Finland and Leningrad has taken an increased interest in the Baltic region of which, as St. Petersburg, the city was historically so much a part. Large sections of the population of Transcaucasia and of Soviet Central Asia in a similar way have a great deal in common ethnically and culturally with the peoples in adjacent regions just beyond the Soviet frontiers. As the peripheral nationalities and regions become economically and demographically more significant, so their importance and influence within the Soviet sphere has become greater, a factor which tends to increase the tendency towards disintegration of the centralised state structure.

As was the case in the period of decline of the dominant states of the past, there have been clear indications of the inadequacy of the Soviet economy to sustain the country's status. As has been observed, the subsequent economic performance did not live up to that initial burst of energy which succeeded in establishing so impressive a heavy industrial base. The enormous resource potential of Siberia proved to be no substitute for technological innovation and more flexible economic strategies. In many ways, the Soviet Union, and especially Siberia itself, had tended by the 1960s to slip into a quasi-colonial position as supplier of energy and raw materials to the more economically advanced parts of the Soviet sphere, many of them lying outside the frontiers of the Soviet Union itself, and also to the countries of the European Maritime Crescent. Bialer sees this as underlying the paradoxical situation of 'external expansion, internal decline',[8] This 'Soviet paradox' he sees as arising from political rigidity, industrial stagnation and economic over-extension into limited fields. However, a paradox it may be, but it is one which is by no means unique to the Soviet Union, since it underlies the conduct of all the dominant states which have been examined. Liska pointed to that 'chasm' which had opened up between the great power of Spain in the late sixteenth century and that 'unproductive economy' which increasingly called into question her capacity to sustain it.[9] This chasm had been bridged for a time by gold and silver from the Americas, until supplies of this began to dry up. He detected 'a similar widening chasm within and beneath the total structure of Soviet power' which for a time, but only for a time, could be bridged by the human and physical resources of Siberia.

The resurgence of dominant state nationalism has also been a feature of the Soviet Union. While the supranational 'Soviet' is the official designation of the state, the idea of 'Russian' has once more risen in importance and esteem. Effectively this dates back as far as World War II when Stalin saw the necessity of using nationalism to help marshal the country's resources and fortify the will to victory. This entailed official endorsement of the central role of the Great Russian nation, a new pride in Russian cultural and scientific achievements, an identification with the Muscovite heritage and a partial rehabilitation of the former Russian national religion of Orthodoxy. Although it was not stated in so many words, this represented an assertion of the hegemonial role of Russia as opposed to the official endorsement of the equality of all the peoples of the Soviet Union. While not officially encouraged, Russian national feeling had persisted and taken forms well in excess of that cultural nationalism which is officially the only form of national expression permissible in a socialist state. Protagonists include the 'Russites', who trace their lineage back to Stalin's official endorsement of the special position of the Russian people. They have taken many of their ideas from the nineteenth century Slavophiles, and they assert the superiority of Russian culture and values.

The incidence of those geopolitical characteristics associated with the decline from dominance is thus a very high one in the Soviet case. Excepting those which concern the actual fall of the dominant state, the Soviet Union has an 83 per cent incidence, which is very close to the average of the other states considered. Next it must be considered to what extent the Soviet Union fits in with those characteristics which concern the actual fall from a dominant position. These are the final bid for supremacy, defeat in war and eventual disintegration into the pre-imperial component parts. As has already been suggested, it can be argued that the final bid for domination has already taken place, and that it took partly a military and partly an ideological form. In military terms, that expansion of the Soviet sphere which took place after World War II made the Soviet Union for the first time overwhelmingly the preponderant power in the western ecumene. However, the European Maritime Crescent, although physically and morally exhausted after six years of war, remained independent of it. The subsequent development of this crescent into an autonomous centre of power took place initially under the Anglo-American military and economic umbrella. This geographical incompleteness was rationalised by Marxists as being an expression of a temporary ideological condition, with an overlapping of systems of production which would be terminated with the inevitable decline of capitalism and the triumph of socialism.[10] With the completion of the transition to a higher form of socio-economic development, this capitalist promontory on the edge of the European land-mass would wither away as an independent socio-economic region, finally becoming what it should be - the western edge of a greater Eurasia working as a unity towards the ultimate goal of Communism. This predicted withering did not take place, and the European Maritime Crescent made a speedy recovery from the war and joined with the United States in the creation of an Atlantic defence and trading system.

Following the final bid for supremacy by a dominant state, further expansionist activity, when it has taken place, has been generally far more low-key, its object being the assertion of a position of limited regional hegemony. Examples of such activity had been that of France in North Africa, and of Turkey in Cyprus and the Aegean. Continued Soviet control over eastern Europe is thus something quite different from an assertion of a wider position of dominance with eastern Europe constituting a kind of springboard for political, ideological and even military control over the European Maritime Crescent. The continued assertion of Soviet control could alternatively be seen as having become no more than the securing of a position of regional hegemony. If this is so, then it becomes a post-dominance phenomenon which accords with the conduct of many of the former dominant states. As to the question of war, in the past wars have usually been the concomitants of decline rather than the initial causes of it. Decline has in fact taken place before the war which finally removes the dominant state and brings about the new international climate which will enable radical restructuring of the political map to be undertaken.

The evidence from this examination of Soviet geopolitical characteristics strongly supports the view that the days of such domination as the Soviet Union has been able to achieve may be coming to an end. The situation fits closely the pattern of decline which has been observed in all the earlier dominating states. The ideological nature of the Soviet bid for supremacy may have been itself unique, but the very existence of the ideology fits in with the pattern of the world-views which have underlain previous bids for supremacy. As has been observed, the nature of all these world-views can be traced back to the original core regions which, after limited territorial expansion, used those human and physical resources at their disposal in an attempt to impose control over the whole of the culture area. In doing this, they were all transformed for a time into missionary states in which the passionate commitment to the spread of the ideology, whatever its nature, took over from more prosaic considerations. However, during the period of decline this total commitment to the ideology begins to waver, and matters of a more practical nature, relating to the maintenance of the position of the state, return into prominence. This is itself related to the geopolitical changes taking place within the state. As it progressively becomes more diverse and centrifugal, and its frontiers become more permeable, so new influences enter and the old certainties begin to be questioned. Eventually the very foundations of the world-view are themselves challenged, and secular pragmatism begins to replace doctrinal certainty.

This may be the underlying significance of the

Gorbachev reforms. While the ostensible purpose of Gorbachev himself and those who think like him is to achieve greater strength through the modernisation of the economy, the implications of glasnost and perestroika[11] could be the recognition of the profound geopolitical changes which have taken place within the country, along with the decision to accept rather than fight against them. If this proves to be the case, then it could herald the peaceful victory of 'the centrifugal view of maritime western Europe' over the centripetal point of view which has characterised Russia. According to Whittlesey, this latter view has tended to take the political form of 'a self-sufficient nation',[12] but in the Russian Empire and subsequently in the Soviet Union it has taken the form of a dominating state, centrally governed, often xenophobic and ultimately seeing the answers to its problems in expansion to encompass more and more geography, which in turn creates more problems. Acceptance of the centrifugal point of view, on the other hand, implies the acceptance of individuality and diversity, with the reduction of central controls to the lowest common denominator of mutual interest and consent. It represents the triumph of the specific over the general, of geography over ideology. Assuming that this is the case, then it could herald the most dramatic geopolitical change to take place east of the Baltic-Black Sea isthmus since Peter the Great brought Russia out of her continental isolation and embarked upon the creation of a world empire.

Is it then realistic to envisage Russia, after so many centuries of dominance being able to accept such a reduced status? The historical evidence drawn from the dominant states of the past is in some ways reassuring. Following the inevitable trauma of decline, they have all been prepared to adapt themselves to the new circumstances of their existence. In fact, the abandonment of pretensions to grandeur has often been followed by economic advance, social reform and cultural revival. Would it be too far-fetched to envisage Russia freed from the burden of her imperial and quasi-imperial commitments, paying more attention to the real well-being of her people and less to chasing the chimera of national greatness and universal mission? This mission after all boils down to the perpetuation of the imported ideas of a German economic thinker which greatly appealed to the late nineteenth century Russian intelligentsia as a way of removing the autocracy. The acceptance of this body of ideas was made easier by the tradition of mutual help and co-operation which goes back to the Russian mir, the old agricultural community. It was to this that the geographer, Prince Kropotkin, had looked back in the early years of the twentieth century as an indigenous model for his ideas on social reform. In his book 'Mutual Aid'[13] he explored the theme of co-operation in human history and found it to have been a far more widespread phenomenon than contemporary Darwinism suggested. He saw the Russian Empire, arrogating all power to itself, as having developed after the manner of Imperial Rome and having crushed the spirit of mutual support among the people.[14] Yet the other side of the Russian character was never entirely stamped out by the crushing power of the imperium. It manifested itself partly in that wild and irrational element in the Russian soul and partly in the values of the narod, the common people, and their astonishing refusal to accept defeat and ability to survive in the face of both natural and man-made disasters. With the decline of the imperium in its contemporary manifestation, the old principles of the mir could appear once more to be a realistic alternative to the wielding of centralised power from above.

NOTES AND REFERENCES

1. R.E. Walters, *The Nuclear Trap: An Escape Route* (Penguin, Harmondsworth, 1974), p. 43.

2. Franz Kafka, *The Castle* (Knopf, New York, 1954), p. 128.

3. Y.M. Goblet, *Political Geography and the World Map* (George Philip, London, 1956), p. 205.

4. Y.M. Goblet, ibid., p. 207.

5. K.A. Wittfogel, *Oriental Despotism. A Comparative Study of Total Power* (Yale University Press, New Haven, 1957), p. 422.

6. Y.M. Goblet, *Political Geography and the World Map*, p. 201.

7. While this applied to the expansion to the west and, to a considerable extent, that to the east, it has not been true of expansion into Central Asia. Although this area lies outside the western ecumene, and consequently outside the strict limit of the present discussion, it does give a clear indication of the Soviet position. The Kara Kum and Kysyl Kum deserts had retarded Russian movement to the south long before they came into contact with the British in the Hindu Kush. Like the British, the Russians then failed to penetrate successfully into the hostile mountain environment of Afghanistan. In the same way, a century later the failure of the Soviet Union to secure control over Afghanistan can be attributed in large part to the problems it faces in an unfamiliar and hostile physical environment. This indicates the expansion of the Soviet sphere being brought to a halt by territorial conditions with which it is unable to cope.

8. S. Bialer, *The Soviet Paradox* (I.B. Tauris, London, 1986).

9. G. Liska, *Russia and World Order* (Johns Hopkins Press, Baltimore, 1980), pp. 74-75.

10. F. Burlatsky, *The Modern State and Politics* (Pro-

gress Publishers, Moscow, 1978), Chapter 4.

11. *Glasnost* is related to the verb *glaset* meaning to announce or declare. It is translated as 'openness', but it also implies honesty in facing up to unpalatable realities and telling the public what these are.

Perestroika means 'restructuring' in the sense of changing some of the bases of Soviet society, and being prepared to accept such changes as necessary.

12. D. Whittlesey, *The Earth and the State. A Study of Political Geography* (Holt, New York, 1939).

13. P.A. Kropotkin, *Mutual Aid* (Heinemann, London, 1902).

14. P.A. Kropotkin, ibid., p. 219.

EPILOGUE

GODS, MEN AND TERRITORY

> Amongst others, I was then encountered, on my passage
> from Westminster to Whitehall, by a tall big gentleman,
> who thrusting me rudely from the wall, and looking over
> his shoulder on me in a scornful manner, said in a
> hoarse voice these words: Geography is better than
> Divinity; and so passed along.[1]

Dr. Peter Heylyn, 'Cosmographia', preface (1649)

Just as there is a subtle relationship between the
physical realties of geography and the ideas which men hold
about what the world is, or should be, about, so there is
also a relationship between the actual process of domination
and the ideas which men have held as to its purpose. To the
Czech writer, Franz Kafka, the 'Silent Castle' embodied his
deepest fears of the state before the power of which the
individual stood helpless and unprotected. Kafka's castle-state
offered no explanations, uttered no sounds on the reasons for
its existence or the source of its authority. It was the
physical expression of power and, as such, its own justification. Another Czech writer, Karel Capek, in his apocryphal
story 'Alexander the Great',[2] has Alexander at the height
of his power pondering on his motives for conquering the
world. He has come to realise that what at the time appeared
to be a heady adventure in which, as a young man, he had
covered himself in glory was, largely unrealised by him at the
time, an enterprise dictated by reasons of state. It was the
threats to Macedonia coming from their neighbours, the
'Thracian barbarians' living in the mountains beyond the
frontiers of Hellenic civilisation, of which his country was a
part, which began that process of expansion which was to
result in the conquest of 'Oriental territories a hundred times
greater than our own country'. In a letter to Aristotle, his
former teacher and respected counsellor, Alexander announces
that he has now decided to become a God. This decision was
also motivated by reasons of state since it would ensure him

160

absolute authority throughout the whole of his dominions and thus enable him in his words, 'to secure for my own land of Greece her natural frontiers on the coast of China. I should thus secure the power and safety of my Macedonia for all eternity'.[3]

In this perceptive tale Capek highlights the two salient features of domination. One is that it is not always what it appears to be at the time, even to those who are themselves most actively engaged in it. There are underlying reasons of state which both initiate the process and then guide the directions which it takes. This suggests a sort of geographical equivalent of Keynes' 'hidden hand', a collective spatial consciousness directing the territorial activity of the state towards what appear to be its best interests. The other is that while expansion may begin as something which is undertaken for specific and rational purposes, it soon gains a dynamic of its own which knows no bounds. As Acton said, power tends to expand indefinitely and will transcend all barriers in doing so. Thus geography is turned on its head and in Alexander's imagination the natural frontiers of Greece become the coasts of China. An attempt is then made to strengthen and justify the whole massive enterprise, rationalising the non-rational, by invoking Divinity in support of geography. Thus the reasons of state of one small country on the edge of the culture area impel it inexorably into converting itself into the universal state of the culture area and then into empire over the whole of the known world.

Throughout the ages these Macedonias - and Castiles, and Prussias and Muscovies - have allowed themselves to be persuaded that they have a special mission, and that their true interests lie in the domination of the world and the creation of the Imperium Orbis Terrarum. In doing so they have come to fit Cellorigo's description of Castile, quoted by Elliott, as 'a republic of bewitched beings, living outside the natural order of things'.[4] However, in modern times not one of them has ever fully succeeded, although many of them have come close to it. The final bid has always proved too costly in men and resources, and the dominant state is eventually replaced by another with fresh ambitions and a new world-view to justify them.

There have been two sides to the condition of dominance as it has affected the inhabitants of the territories dominated. The positive one is that it has brought order to the culture area by reducing internecine conflicts within it and establishing an overall system. In this way the Imperium produced the pax Romana, the pax Turcica and the pax Russica, and within this 'imperial peace' human energies could be devoted to more positive matters than warfare. The negative side to this is that the power of the dominant state within its territories is largely unbridled. It is the state which itself determines the nature of this power and how it shall be exercised. Its twin bases are the Castle and the God, the former the physical expression of the existence of power and the latter the higher justification for the exercise of it. This is Divinity harnessed in support of geography, and the anthropocentric world-view in support of the acquisition of the world. The two are closely linked since the particular shape of the Divinity was forged in those same territorial conditions in which the will to dominate originally arose. The acceptance of this Divinity, expressed in modern secular terms as ideology, is the tribute which has to be paid to the dominant state by those whom it dominates. The price has proved to be a very high one entailing the relinquishment by the subject peoples of many freedoms, both material and spiritual. Alongside this has come a tendency towards entropy, a slowing down of cultural, social and economic activity which has often had the effect of converting the 'imperial peace' into an 'imperial backwardness'. This includes that 'stagnation, epigonism and retrogression' identified by Wittfogel, unpleasant side-effects of that peace and order which the dominant state bestows.[5]

Since World War II attempts have been made to establish peace and order within the western ecumene without recourse to that domination which in the past has been the inevitable concomitant of it. In the countries of the European Maritime Crescent an attempt has been made to establish by consent something closer to the Duc de Sully's grand dessein than to Le Roi Soleil's arrogant domination. The as yet limited and functional structures developed by the European Communities, which now include in their membership three of those powers which made bids for supremacy over the continent, have laid the foundations for a different type of international order from that which in the past has been associated with dominance. While its objective too is peace and order, it represents a fundamentally different approach to the problem of achieving it. It is founded on what Whittlesey saw as being the 'centrifugal view' of maritime western Europe, and this puts it into the lineage of the res publica rather than the imperium.

The tradition of the res publica has persisted through the ages in the maritime states, free cities, republics, small nationalities and mountain confederations located mainly in the western and central regions of Europe. To the east such small units have not normally been able to survive the onslaughts of the larger ones. They have easily been absorbed into that imperium which has been considered to be the normal state of affairs. The devolved, the centrifugal, the small unit has not been much part of the historical experience of those Europeans living to the east of the Baltic-Adriatic isthmus. yet the unscrambling of the contemporary eastern imperium is as much a prerequisite of the centrifugal approach to international order as was the curtailment of the unbridled aspirations of the great powers in the west of the continent.

The Russians, said Berdyaev, are an apocalyptic people. They are not satisfied with western-style compromise and are passionately desirous of achieving ultimate solutions to the gigantic problems arising from their human and physical conditions.[6] The solution which presented itself during the first quarter of the twentieth century was the Revolution, but during the second quarter this led back via the imperial road to centralisation and the perpetuation of the condition of dominance. The solution which Kropotkin had proposed, fifteen years before the Revolution, involved a return to the 'mutual aid' of the Russian mir. Kropotkin, the anarchist geographer, went behind ideology, behind the grandiose world-views which have been the traditional concomitants of dominance, to what he perceived to be the necessities dictated by geographical reality.[7] Co-operation, he asserted, is not simply something laudable and desirable; it is an essential element of behaviour if mankind is to survive in the world. Kropotkin's message comes through with even greater urgency today in the age of Mutually Assured Destruction (MAD) than it did in those calmer days at the beginning of the century. Likewise, the replacement of imperium by res publica in international relations may also prove to be not merely a worthy objective, but a necessity for human survival. As has been noted, Louis Janz saw the idea of unity in the past as having been inevitably associated with 'expansion, annexation, conquest, Anschluss, penetration, occupation and protectorate'.[8] Other expansionist states may now be waiting in the international wings for the moment when they consider it propitious to make their own bids for supremacy. Aggressive behaviour by certain states in the Mediterranean and the Middle East suggests the possibility of the existence there of putative dominating states. Such states could seek to use the opportunity afforded by international instability to increase their power and influence and to establish a base from which territorial domination could be attempted. However, it appears highly unlikely that military supremacy of the sort which has been examined in this book could ever be achieved and retained in the nuclear age. The damage done to the whole human fabric of the western ecumene and beyond, and therefore to that of the expansionist state itself, would in all probability prove to be irretrievable. It follows that the creation of order by means of territorial domination thus appears to be no longer a realistic option. Yet, even though it may appear unrealistic to the reasonable mind, it is by no means beyond the bounds of possibility that it may be attempted by certain states which have become imbued with a sense of messianism and destiny. When the 'bewitched beings' take over, then realism and rationality cease to be prime considerations, but the consequences of their actions could be far more catastrophic than at any previous time in history. It is no longer sufficient to ask, as did Alexander Blok's Scythians

> Are we to blame if your rib-cages burst
> beneath our paws' impulsive ardour?[9]

Such ardour moulded the geopolitical world in the past, but it may destroy that of the future unless it is actively prevented from doing so.

Thus it becomes all the more essential that the protection of today's Macedonians from today's Thracians should take a form totally different from that which appeared the best, indeed the only one, to Capek's Alexander the Great. Rather than rulers seeking to become gods - or their contemporary equivalents - and 'natural frontiers' being extended to the coasts of China, it is necessary in the interests of survival that Macedonians and Thracians be brought together in a structure of peace and stability within the wider region of which they both form a part. The geopolitical viewpoint, and the insights into territorial motivations which its study gives, can be of considerable value in coming to terms with those turbulent forces which underlie the world political map. By attempting to understand them, rather than by trying to pretend that they do not exist, it may be possible to divert them towards co-operative rather than confrontational solutions to the problems which the world faces. The creation of a stable political surface could be the greatest challenge which either the apocalyptic Russians or the pragmatic westerners have yet had to face. If successful, it could be the beginning of a more tranquil chapter in the long and turbulent history of the western ecumene and indeed of the whole world.

NOTES AND REFERENCES

1. Prefatory quotation from S.W. Wooldridge and W.G. East, *The Spirit and Purpose of Geography* (Hutchinson, London, 1951).

2. K. Capek, *Apocryphal Stories*, trans. D. Round (Penguin, Harmondsworth, 1975).

3. K. Capek, ibid., p.36.

4. J.H. Elliott, 'The Decline of Spain' in C.M. Cipolla (ed.), *The Economic Decline of Empires* (Methuen, London, 1970), p. 195.

5. K.A. Wittfogel, *Oriental Despotism. A Comparative Study of Total Power* (Yale University Press, New Haven, 1957), p. 422.

6. E. Lampert, *Nicolas Berdyaev and the New Middle Ages* (James Clarke, London, n.d.), p. 73.

7.   P.A. Kropotkin, *Mutual Aid* (Heinemann, London, 1902).

8.   L.   Janz,   'The   Enlargement   of   the   European Community', *European Community*, No.1, 1973 (European Community Office, London)

9.   A. Blok, *The Twelve and Other Poems*, trans. Jon Stallworthy   and   Peter   France   (Eyre   and   Spottiswoode, London, 1970), p. 163.

GLOSSARY OF GEOPOLITICAL TERMS USED IN THIS BOOK

Anschluss   annexation of a country, or part of one, by another country. The word was used to describe the union of Austria with Nazi Germany following the occupation of Austria by the Wehrmacht in 1938. The word was then used by certain French political geographers to denote the process of piecemeal take-over which characterised the expansion of Prussia and subsequently of Germany.

Ausgleich   settlement or compromise. It refers in particular to a change in the Austrian constitution in 1867 which brought about a large measure of autonomy in the government of Hungary and resulted in considerable Hungarian participation in the running of what was from then on the Austro-Hungarian Empire.

Blitzkrieg   'lightning war' based on the theory of the best use of tanks and other armour in land warfare. It was put into practice by the Wehrmacht in World War II and resulted in an astonishing series of victories from 1939 to 1941.

Cold War   the situation of confrontation and antipathy between the western and the Communist blocs which became a central feature of their relations following the end of World War II. It is especially associated with that exceptionally bleak period between the proclamation of the Truman Doctrine in 1947 and the death of Stalin in 1953.

Conditio Germaniae   an adaptation from the descriptive term Conditio Preussen. This refers to the fate of Prussia to be located in the heart of Europe, formerly the continent's principal battleground and subsequently surrounded by often hostile great powers. This position was used as the justification for the adoption of ruthless measures in order to ensure the survival of the state. It was a condition which was seen by some German politicians and political thinkers as having been inherited by the German Empire.

Cordon sanitaire a zone established around the frontiers of a particular state and designed to prevent or discourage further expansionist activity, or a zone between two states designed to keep them apart. A zone of this sort is normally made up of a number of small states which themselves have a vested interest in the maintenance of peace and which are encouraged to maintain a neutralist or independent stance.

Core area the nucleus or central region of the state. The term has come to be used in a general way as though it always meant the same thing. In fact a number of different types of core areas can be distinguished within the same state, each of which has had a different function within the power structure.

See also Original core area, Historic core area, Economic core area.

Core nation the homogeneous population group which is central to the expanding state. It is the language and culture of this nation which become those of the dominant state as a whole.

Culture area the geographical area occupied by the culture.

See also Parent culture.

Deutschtum term used to denote the area occupied either wholly or in part by the German people in Europe.

Drang nach dem Süden German longing or yearning for the south expressed in the attempt to be associated with the Mediterranean world and in particular with Italy. This was very much a characteristic of certain periods of the Holy Roman Empire and it resurfaced in the nineteenth century with the acquisition by Austria of considerable territory in northern Italy.

See Chapter 3 n 48.

Drang nach Osten the historic German urge to expand towards the east. The marks, or marcher regions, on the eastern frontiers of the Holy Roman Empire were the original bases for this expansion which eventually brought vast territories under the control of the German-speaking countries.

Economic core area this constitutes the principal centre of economic power within the state. Its location will change as geographical values change consequent upon technological advances and the harnessing of new physical resources. At different periods and in different states agriculture, commerce, finance, manufacturing and heavy industry have been the principal bases of strength, both political and military.

Ecumene the most populous part and developed areas of any country or landmass. It stands in contrast with those areas which, as a result of their relatively less favourable physical conditions, are less able to support intensive development.

See Chapter 1 n 3.

European Maritime Crescent this forms a gigantic swathe around the European continent from Scandinavia in the north-east to Turkey in the south-east. It consists mainly of islands and peninsulas linked together by the Franco-German quadrilateral at its centre. With the exception of those countries lying around the Baltic Sea and the Anatolian plateau, it lies entirely outside the Heartland (q.v.). It is this, together with the strength of maritime influences, which gives this extremely diverse area its geopolitical character. Since World War II the countries of the Crescent have been moving more closely together in both defence and economic matters.

Festung Europa fortress Europe. With the turning of the tide in World War II, German-occupied Europe began to come under attack from the east, south and west. The Nazi concept of an embattled European fortress was intended to inject strength and purpose into the defence. Among its manifestations were the Westwall (Atlantic Wall), and later the idea of an Alpine redoubt.

Geopolitical centre this concept brings together the original core area (q.v.) of the state together with its historic core area (q.v.) and the territory over which the core nation (q.v.) first secured control before expanding out to establish a wider position of dominance.

Geopolitics this is the study of states as spatial phenomena with a view to reaching an understanding of the geographical bases of their power. State behaviour is examined against the background of such characteristics as territory, location, resources, population distribution, economic activity and political structure. Each state is examined as a component part of the world's political space, and the consequent pattern of international relationships constitutes an essential part of the study. Geopolitics is thus holistic in its approach, the object being to bring together diverse phenomena and to describe and interpret them as a totality. Contemporary geopolitics is in no way to be confused with German Geopolitik (q.v.) as it existed during the Nazi period.

Geopolitik term attributed to Rudolf Kjellen (1846-1922) who defined it as the science which conceives of the state as a geographical organism and phenomenon in space. In Nazi Germany it came to be perverted into an instrument of state policy and was used as a justification for German pre-eminence in Europe.

Grossdeutschland 'greater Germany'. This was originally thought of as the outcome of the union of Germany (Kleindeutschland q.v.) with Austria, but later came to be conceived of as being the union of all the German peoples in Europe. The expansion of the Third Reich beginning with the Anschluss (q.v.) of 1938 briefly converted the idea into a reality.

Heartland term used by Halford Mackinder in 1919. It

was essentially a modification and adaptation of the Pivot (q.v.) to fit the conditions at the end of World War I. It covered a considerably larger area than the Pivot by loosening the criteria for its delimitation. It was extended westwards to include the drainage basins of the Baltic and Black Seas which Mackinder saw as being out of the range of sea power. It thus extended across eastern Europe to the Elbe and included most of what is now the Soviet sphere.

Herrenvolk the idea, particularly favoured by the Nazis, that the Germans were a master race and thus superior to all except the other 'Aryans'. It followed from this in their thinking that the Germans were the natural leaders of Europe, while 'inferior' races, such as Jews and Slavs, would thus occupy a lowly position in the New Order (q.v.).

Historic core area the state established as a result of territorial expansion from the original core area (q.v.). Its political position is that of a march, or mark, within the territory of its parent state (q.v.). It is organised on a military basis and is characterised by a propensity for aggressive and expansionist activity. Such activity will make it increasingly autonomous, and eventually it is likely to become politically independent of the parent state.

Inner Crescent term used by Halford Mackinder for those lands lying around the edges of the Heartland (q.v.). They included the whole of continental Europe, the Middle East, the Indian sub-continent and China.

Iron Curtain see Chapter 6 n.6.

Irredenta unredeemed territories. The term is derived from Italia Irredenta which was used by Italian nationalists, including the Fascists, for what they considered should be the real territory of Italy. After the Treaty of Versailles the Italian frontiers still fell far short of this, and a considerable sense of grievance was felt on that account. The term irredenta has subsequently been used to denote those areas which states feel should, for a variety of cultural and historical reasons, belong to them. Such territories constitute a part of their geographical self-image, and they may use any opportunities offered by international instability to regain them.

Italia Irredenta see Irredenta

Kleindeutschland 'little Germany'. This was created by the union of Prussia together with the south German states into the German Empire in 1871. It had been Prussia's policy to exclude Austria from this union since Austria, the old hegemonial power of the Holy Roman Empire, was considered at the time to be a rival to unbridled Prussian dominance. Later a movement began, both in Germany and Austria, to bring about the union of the two countries.
See Grossdeutschland

Kulturboden term used in German Geopolitik to describe that area in which German civilisation (Kultur) was considered to be dominant.

Lebensraum literally 'living space'. The term was used particularly by German political scientists and by the Geopolitiker to indicate the territory which it was considered Germany required to support her growing population and to maintain her status as a great power.

Limites naturelles the French geopolitical concept of natural frontiers. According to this every state has a 'natural' territory to which it is entitled to aspire, but beyond which it may be dangerous to attempt to expand. The idea has been current since the seventeenth century when it was used to justify the extension of France eastwards as far as the Rhine. In recent times it has been criticised by many French geographers both on account of its inflexibility and for ignoring the dynamic relationship between man and his environment.

Lotharingian axis see Chapter 2 n 17.

Macrocore the principal core region of the culture area (q.v.). This region is the focus of commerce, industry and the communications network. It is likely also to contain the major political, religious and cultural centres.

Mare Nostrum 'our sea'. A term used by the Romans for the Mediterranean Sea in order to indicate their sense of possession and domination. In the 1920s as 'Mare Nostro' the term was resurrected by the Italian Fascists. Like the use of the fasces itself, it was intended to associate Fascist Italy with the Roman Empire and to indicate the aspiration to achieve a similar commanding position throughout the region. See Chapter 1 n 25.

Mitteleuropa central Europe. Because of its indeterminacy this has proved to be a difficult geopolitical concept to delimit with any degree of precision. While it has quite clear boundaries to the north and south - the Baltic Sea and the Alps - the major physiographic trend of the continent is from west to east, and so it tends to merge with other macroregions. Much attention has been given by German geographers to the establishment of a set of criteria as a basis for more precise delimitation. While basically a geographical concept, it was politicised when it came to be associated with those areas in which Germany and the German people were in a commanding position.

Musspreussen forced Prussianisation of the other German states associated with Prussia's increased westerly orientation after 1815. This process culminated in 1871 with the establishment of the German Empire under Prussian hegemony with the subsequent attempt to introduce Prussian methods and values in the west and south of the new state.

Naturgrenzen natural frontiers. The difficulty of finding such frontiers in the ill-defined physical conditions of Mitteleuropa (q.v.) led to the German attempts to define the limits of the German nation in cultural rather than physical terms.

See also Limites naturelles

Neuordnung   See New Order

New Order   Die Neuordnung was a Nazi phrase to describe the new European system which they aimed to establish. It was to be organised and led by the Herrenvolk (q.v.) and the old states were to be abolished in favour of client states grouped around Grossdeutschland (q.v.).

Original core area   the political nucleus out of which the state subsequently grew. In the states examined in this book the area lay in the frontier regions of a parent state and was established by the rulers of the latter in order to stabilise the frontiers and act as a base for further expansionist activity.

See also Historic core area

Ostrog   fort or fortress constructed of timber. Such forts became the centres of Russian political and military power as they moved eastwards across Siberia and subjected the native peoples.

Parent culture   the overall culture to which a particular state belongs. In this sense 'culture' is defined as being an interlocking system of human cultural relationships which is based upon a particular religious or secular world-view (q.v.).

Pivot   term used by Halford Mackinder in 1904 to denote that area in the centre of Asia having inland or Arctic drainage. It was consequently out of reach of maritime power and Mackinder saw it as being the key geopolitical area of the world.

See p. 96.

Realpolitik   political action based upon an evaluation of world realities rather than upon idealistic notions. In international relations it implies an acceptance that force may be necessary in order to achieve objectives.

Sprachboden   the geographical area in which a particular language is spoken. Term used in German Geopolitik to indicate the area within which German was either the only or the major language spoken.

See also Kulturboden and Volksboden.

Staatsidee   the state idea. It implies a philosophical and moral concept of the destiny and mission of the state in the world.

See World-view.

Volksboden   the geographical area inhabited by a particular people (Volk). Used in German Geopolitik to indicate the total area inhabited by the German people in Europe.

See also Kulturboden and Sprachboden.

Weltpolitik   world politics in the sense of the advocacy of a world role for Germany in the first half of the twentieth century. This was based on that school of thought which placed great store by the acquisition of colonial territories and the enlargement of the German fleet so that Great Britain's world power could be challenged. The advocates of this strategy were opposed by those who favoured a policy of Lebensraum (q.v.) although the two strategies were not inevitably contradictory.

Western ecumene   See Chapter 1 n 3.

World-Island   See Chapter 1 n 3.

World-view   term used in this book to denote the central idea which lies behind the expansion which leads to dominance. It is used by the dominant state as its raison d'etre and as the justification for its dominant position. Such world-views may take a religious or a secular character.

INDEX

Figures in brackets refer to note numbers
F = Figure; T = Table

Acton, Lord 3, 7, 8, 10(8), 161
Adrianople 13
Adriatic Sea 30, 32; region 105, 108
Afghanistan 137, 158(7)
Africa, North 21, 59, 156
agriculture 14, 16, 20, 24, 38, 39, 44-5, 47, 55. 78. 89. 114
Albania 108, 123
Alexander the Great 160-1, 164
Alps 8, 28, 30, 37-8, 40, 44, 55, 76
Alsace-Lorraine 49, 52, 58, 152
Altmark 46, 47, 64, 68
America 21-3, 24, 25-6, 28, 154; see also United States of America
Amur, R. 116
Anatolia 12, 14, 15, 16-17, 19, 93, 127, 151
Ancel J. 40, 41, 45, 62(11)
Andalusia 20, 24, 152
Anjou 38, 39
Ankara, Battle of (1402) 13
Anschluss 7, 163, 166; Austrian (1938) 52; 'Rhénane' 58
Arab Empire 17-18, 19, 20; see also Islam
Aragon 20, 24, 25, 68
Arctic Ocean 82, 83, 89-90, 96, 103, 126

Arendt, H. 124, 131(24)
Armada, see Spanish Armada
Armenia 112, T2, 149
Asia 7, 12, 16, 76, 86, 141; Soviet, see Aziatchina
Astrakhan 78, 82
Atlantic Alliance 155-6
Atlantic Ocean 27, 38, 43, 53
Attila the Hun 7
Auerstadt, Battle of (1806) 43
Augustus, Emperor 4, 8
Ausgleich (1867) 36, 166
Austerlitz, Battle of (1805) 43
Austria 47, 49, 51, 52, 55, 56, 61(6), 80, 154
Austrian Empire 30-36, F3.1, 39, 69, 85, 149, 151; development 31-2, 33-5; dominant state character-istics 64, 65-8, T1, 70, 90, 151, 152; fall 35-6, 113, 151; geographical background 30-1, 32-3; origins 30, 64
Austrian Peace Treaty (1955) 108
Austro-Hungarian Empire, see Austrian Empire
Azerbaijan 112, T2
'aziatchina' 134, 144(13); see also 'evraziistvo'
Azov, Sea of 80

Baikal, L. 114, 116, 131(14)
Baku 53
Bakunin, M. 85

balance of power 2-3, 8
Balearic Islands 20
Balkans 13, 14, 15, 16, 18,
   31, 33, 52, 55, 68, 85, 93,
   105, 108, 122, 148, 152
Baltic Republics 53, 55, 80,
   86, 103, 104, 116, 153; see
   also Lithuania
Baltic Sea 30, 31, 46, 47, 52,
   53, 55, 77, 80, 81, 82, 83,
   85, 87, 90, 91, 94, 96,
   103, 105, 108, 117, 118,
   119, 121, 126, 157
Baltiiysk (Pillau) 118
Barbarossa, Operation 53,
   59-60, 104
Barbary corsairs 13, 18
Barber, N. 28-9(13)
Barcelona 24
Barghoorn, F.C. 130(9)
Bassin, M. 63(39)
Bater, J.H. 91, 100(32)
Bavaria, Duchy of 30
Berdyaev, N. 82, 99(6), 124,
   131(27), 163
Berlin 46, 51, 69, 104, 128
Berlin-Baghdad railway 55
Bessarabia 80, 103
Bevin, Ernest 124
Bialer, S. 154, 158(8)
Bielfeld, J.F. von 62(12)
Bismarck, Otto von 6, 49, 148
Black Sea 52, 53, 76, 77, 80,
   81, 82, 83, 84, 85, 90, 91,
   93, 96, 103, 108, 116, 126,
   153, 157
Blackstock, P.W. 131(17)
Blij, H.J.de 62(25)
Blitzkrieg 52-3, 59, 166
Blok, Alexander 142-3,
   146(36), 164
Bohemia 52
Bolshevik Revolution (1917)
   109, 113, 121, 123-4, 125,
   134, 139, 153, 163
Bosnia 36
Bosphorus Sea 122-3, 124
Bowman, I. 84, 100(13)
Brandenburg 46, 51, 52, 118

Braudel, F. 3-4, 6, 10(9),
   10(17), 14, 16, 18, 24-5,
   27, 28, 30, 35, 68
Brazil 140
Bremen 62(26)
Brest-Litovsk, Treaty of
   (1918) 52, 103, 104, 111
Britain, see Great Britain
British Empire, see Great
   Britain
Brown, N. 137, 140,
   144-5(21)
Brüderkrieg 49
Bryce, J. 6, 10(19)
Bucharest 108
Bug, R. 121
Bukharin, N. 134, 144(13)
Bulgaria 84
Bullock, A. 131(22)
Burgos 20, 29(18)
Burgundy 38
Burlatsky, F. 158-9(10)
Bursa 12
Byelorussia 53, 92, 103, 104,
   112, T2
Byzantine Empire 5, 12-13,
   14, 16, 17, 18, 64, 69, 77,
   78, 83, 84, 90, 93, 122,
   123-4, 139, 141
Byzantium, see Constantinople

Cadiz 26
Cajal R.y 23-4
Callander, T. 60, 63(50)
Calleo, D. 63(40)
Callipolis (Gallipoli) 13
Canada 140
Cantabrian Mts. 19, 20
Capek, K. 160-61, 164,
   164(2)
capital cities, role of 13, 17,
   30, 39-40, 69-70, 73, 76,
   80, 82-3, 85-6, 92, 111-12,
   117, 119, 124-5
Carinthia 31
Carlowitz, Treaty of (1699) 32
Carniola 31
Carnot, L. 41
Carpathian Mts. 32, 84, 93,
   105, 118, 151

Caspian Sea, 16, 53, 76, 78-80, 81, 82, 96
Castile 20, 23-5, 27, 28, 29(18), 64, 92, 161
Catherine II (Catherine the Great) 85, 87-8
Catholicism 5, 18, 20, 22-3, 24, 30-32, 45, 68, 83; see also Holy Roman Empire
Caucasus Mts. 53, 81, 84, 91, 93, 97, 104
Cellorigo 161
Charlemagne 5, 20, 36, 38, 65, 68, 69
Charles V (Holy Roman Emperor) 8, 21, 31
Charles VIII of France 39
Charques, R. 99(3)
Chessin, S.de 133-4, 137, 143(2)
China 127, 140, 141; Chinese empire 92, 98
Christendom, see Christianity, Holy Roman Empire
Christianity 5, 9, 12, 16, 17, 18, 19, 20-3, 30-2, 38, 45-6, 59, 64, 69, 81, 83-4, 85; see also Holy Roman Empire, Russian Orthodox Church
Churchill, W.S. 105, 130(6)
Cicero 7
Cisleithania 33-4, 36
Clauswitz, K. von 51
climate 19, 43, 59, 60, 87, 92, 98, 148-9
Cold War 166
Columbus, Christopher 21-3
Comecon (Council for Mutual Economic Assistance) 108
Cominform (1947) 108
commerce 14, 24, 25, 31, 35, 38, 40, 43, 44, 57, 70, 77, 78, 80-81, 82-3, 84, 89, 91
communications 26, 32-3, 38, 55, 57, 64, 70, 76, 77, 82-3, 86-7, 116, 151
Communism, see Marxism
'conditio Germaniae' 166
'conditio Preussin' 50, 166

Confederation of the Rhine 42
Congress of Vienna (1814-15) 42, 49
'Congress System' 6
Constantinople (Byzantium) 12, 13, 17, 77, 84, 85, 91
Cordoba 20
Core area 12-13, 16, 17, 27, 64-75, T1, F4.1, F4.2, F4.3, 78, 81-2, 89, 122-3, 149-52, 167; economic 24, 51, 167; historic 23, 30, 33, 45-6, 49, 65, T1, 73, 97-8, 149-51, 169; original 64-5, 171
Core nation 65-6, 70-4, 167
Cornish, V. 28(1), 30
Cossaks 78, 82
Council for Mutual Economic Assistance see Comecon
Cressey, G.B. 137, 145(24)
Crimean War 97
Culture 1, 2, 39-40, 55, 56-7, 67-8, 70-4, 167
Curzon Line 103-4, 117, 130(1)
Czechoslovakia 52, 103, 105, 118, 152

Danilevski, N.J. 100(15)
Danton, G. 41
Danube, R 13, 17, 18, 30, 32-3, 35-6, 50, 55, 68, 84, 85, 105, 118, 151
Danzig 47
Dardanelles 13, 123, 127
Darwinism, social 56, 57, 68, 157-8
demography, see population
Denmark 49
Despicht, N. 29(17)
Deutsches Reich, see German Empire, Third Reich
Deutschtum 57, 58, 65, 167
Dewdney, J.C. 28(4)
Dickinson, R.E. 63(37)
Diettrich, S deR. 100(23)
discoveries 21, 27
Dniepr, R. 77, 78, 80, 82, 83
Dniestr, R. 91

dominance, geopolitical model of 64-75, T1, F4.1, F4.2, F4.3, 89-91, 93-4, 118-22, 126-7, 129, 135; retreat from T4, 152-7
'dominium' 4, 8
Don, R. 77, 80, 82, 91, 103
Donbas 114, 152
Douro, R. 19
'Drang nach dem Süden' 83, 166; see also 'Urge to the Sea'
'Drang nach Osten' 55, 82, 103, 167
'Drang nach Westen' 82, 103
Dutch War of Independence 25
Dvina, R 80

East, W. G. 135, 144(17), 164(1)
East Germany, see German Democratic Republic
East Prussia 104-5, 118
Ebro, R. 19
economic strength 12-13, 16, 17, 24, 25-8, 31, 35, 39, 43-4, 47, 49-50, 51, 55, 56, 57, 58, 68, 69-70, 82, 89, 93, 97, 109, 113-14, 119, 121, 128, T3, 136-7, 149-51, 153
ecumene 9-10(3), 167; Western 1, 8, 9-10(3), F7.1
E.E.C., see European Economic Community
Egypt 13
Elbe, R. 42, 46, 47, 49, 105, 118
Elliott, J.H. 25, 29(23), 161
empire, see imperialism
Enlightenment 40, 45, 68, 69
Escorial, Monasterio del 24, 28, 149
Estonia 80, 103, T2, 154
Eurasia 76, 82, 91, 137
Eurasianism in USSR, see 'Evraziistvo'
European Economic Community (EEC) 127-8, 162

European Maritime Crescent 125, 155-6, 168
'Evraziistvo' 86, 98-9, 115-17, 125-7, 128-9, 134, 137, 141, 142-3, 158(7)
expansionism 1-6, 8-9, 147-8, 160-4; see also individual empires

Falkus, M.E. 131(23)
Far East 86, 88, 91, 96, 141; see also China, Japan
Fascism 10, 11(24); see also National Socialism
Fawcett, C.B. 135
Federal German Republic 130(7), T3, 154
Ferdinand II of Aragon 20, 21
'Festung Europa' 60, 168
Finland 80, 103, 118, 154; Gulf of 46, 104-5, 117-18
Fisher, H.A.L. 31, 61(3)
Five Year Plans (USSR) 113-14, 117
Florence 21, 42
Forbes, N. 85, 92, 100(16)
France 26, 50, 51-2, 53, 54-5, 103, 135-7, T3, 140, 156; see also French Empire
Francis II (Holy Roman Emperor) 32
Frankfurt Parliament (1848) 50
Frankish empire 65, 68; see also Charlemagne
Franz Ferdinand, Archduke 36
Frederick the Great 47, 60
French Empire 36-45, F3.2, 47-9, 54-5, 57-8, 76, 90, 92, 97, 118, 142; development 36, 39-42; dominant state characteristics 65, T1, 68, 69-70, 151, 152; fall 42-4, 59, 97, 98, 151-2; geographical background 36-8; origin 36-39, 65
French Revolution 32, 41-2, 44, 45, 50, 152

frontiers 2-3, 14, 40-1, 45,
47-8, 50, 53-5, 57-8, 68,
69, 72-3, 82, 83, 89-90,
F6.1, F6.2, 122; see also
'Limites naturelles',
'Naturgrenzen'

Galicia 118
Gallia (Gaul) 38, 65
'Gemeinschaftsgefuhl' 57
Gengis Khan 7, 77-8, 133
Genoa 13, 26
'Geopolitik' 56-7, 58-9, 60,
168
Georgia 81, 112, T2
German Democratic Republic
118, 128, 130(7), 154
German Empire (Second
Reich) 36, 45-52, F3.3,
81, 85, 90, 93, 103, 121,
140-1, 142, 161, 166;
development 49-51, 64-5;
dominant state character-
istics T1, 69, 70,
144-5(21); fall 51-2, 148-9;
geographical background
46-7; origins 45-9; see
also Holy Roman Empire,
Third Reich
Germany, see German Empire,
Holy Roman Empire, Third
Reich; East, see German
Democratic Republic; West,
see Federal German
Republic
ghaza, see Holy War
Gilbert, E.W. 135, 144(16)
Glasnost 157, 159(11)
Glassner, M.I. 62(25)
Goblet Y-M. 2, 10(4), 147,
149, 152
Goebbels, Joseph 59, 63(46),
130(6)
Golden-age myth 2, 147-8
Gorbachev, M. 156-7
Gottmann, J. 126, 130(8)
Gould, P.R. 9
Granada 20, 21
'grand dessein' see Sully,
Duc de
Gravier, J.F. 61(9)

Great Britain 26, 35, 39,
42-3, 44, 47, 50, 51, 53,
55, 56, 59, 60, 76, 85, 97,
98, 99, 103, 123, 135-7,
T3, 151, 155; British
Empire 56, 99
Greece 13, 18, 93
Gregory, D. 11(26)
Grenier, A. 6
'Grossdeutschland' 52, 60, 168
Guadalquivir, R. 19, 20, 24,
26, 151
Gurian, W. 130(2)

Habsburg dynasty 21, 31, 32,
33, 35, 36, 68, 149
Halecki, O. 96, 101(36)
Hamburg 62(26)
Hanseatic League 31
Hatton, R.M. 86, 100(21)
Haushofer, K. 56, 63(43)
Heartland theory (Mackinder)
134-5, 137, 140-1, 144(15),
169
Henrikson, A.K. 1, 2, 9(2),
139
Henry I, Duke of Saxony 46,
62(21)
Hercynian mountains 36, 76,
93
Herzen, A. 86
Heylyn, P. 160, 164(1)
Hitler, Adolf 57, 59, 60, 148;
see also National Socialism
Hohenzollern dynasty 46
Holland, Kingdom of 42; see
also Netherlands
Holstein 49
Holy Alliance 21
Holy League 18
Holy Roman Empire 5, 13, 18,
21, 30-32, 33, 38, 42,
45-6, 50, 55, 62(21), 64
Holy War, Islamic 12, 14
Hooson, D.J.M. 116,
130-31(13)
Hoselitz, B.F. 131(17)
Hunczak, T. 86, 87, 100(14)
Hundred Years War 39

Hungary 13, 31, 32, 33, 35-6, 52, 55, 61, 65, 149, 154; see also Austrian Empire
Huttenbach, H.R. 82, 83-4, 94, 99(2)(5)

Iberia, see Portugal, Spain
ideology 4, 9, 44-5, 68, 69, 75, 76, 83-4, 86, 110, 119, 123-4, 156, 162; see also Communism, religion
Ile de France 38
Imperial peace 6-7, 161
imperialism 3-9, 20-1, 56, 60-1, 86, 147-58, 162; see also individual empires
Inalcik, H. 28(10)
India 92, 140, 141
Industrial Revolution 44
industry, see economic strength
Ingria 80
Iran 137
Iron Curtain 105, 123, 130(6)
'Irredenta' 169
Isabella of Castile 20, 21
Islam 9, 12, 13, 14, 16, 17-18, 19-20, 27, 31-2, 64, 68, 81; see also Ottoman Empire
Italy 10-11(24)(25), 13, 17, 21, 25, 26, 32, 35, 39, 42, 44, 53, 55-6, 57, 60, 94, 142; see also Roman Empire
Ivan III, Tsar 78
Ivan IV, the Terrible 78, 87

Janissaries (Yeni Ceri) 16
Janz, L. 7, 10(23), 163
Japan 88, 96, 98, 141; Sea of 117
Jena, Battle of (1806) 43, 47
Jews 24, 57, 59, 149
Joan of Arc 39
Jones E.L. 62(24)
Jones, R. 28(2)
Jones, S.B. 135, 144(19)
Juel, Just 88
Julius Caesar 7, 41

Kafka, Franz 135, 147, 158(2), 160
Kaliningrad (Königsberg) 118
Karaganda coalfield 114
Karelia 80, 104
Kazakhstan 112, 114, T2, 116
Kazan 78, 82
Kennan, G. 124
Kerner, R.J. 76, 99(1)
Keynes, J.M. 161
Khruschev, N. 108
Kiev 77, 78, 83, 125; Kievan Russ 77-8, 80, 82, 83, 89, 91, 118, 122, 139
Kirby, S. 137, 142, 145(23)
Kirgizia 112, T2
Kjellen, R. 56, 63(35)
'Kleindeutschland' 49, 50, 52, 55, 63(48), 65, 92, 169
'Kleinstaatengerumpel' 59, 63(46)
Kolarz, W. 40, 61(10), 85
Königgrätz-Sadowa, Battle of (1866) 36
Kossovo Polje, Battle of (1389) 13
Kristof, L.K.D. 2, 10(5), 86, 87, 112, 126, 127, 132(35), 134, 139, 141
Krivoi-Rog 152
Kropotkin, P. 157-8, 159(13), 163
Krosig, S.Von 130(6)
'Kulturboden' 57, 60, 170
'Kulturgrenze' 55, 56-7
Kuzbas (Kusnetsk Basin) 114, 116-117, 153

Ladoga, L. 104-5, 117-18
Lampert, E. 164(6)
language 17, 39, 50, 57, 85, 89, 112
Latvia 103, T2, 116, 131(14)
Lavisse, E. 134, 144(11), 153
leadership 1, 147-8, 164
League of Nations 1
'Lebensraum' 55, 56, 57, 58-9, 60, 170
Leitha, R. 33
Leitsch, W. 99(9)

Lenin, V.I. 133; Leninism
102, 108, 109, 112, 113,
124, 125, 127, 128, 133
Leningrad 53, 104-5, 113,
116, 121, 125, 154; as
Petrograd 103, 119; as St.
Petersburg 80, 85-6, 90,
91, 92, 111-12, 125,
131(14)
Leon, Kingdom of 29(18)
Lepanto, Battle of (1571) 18,
21
Libya 123
Lichtheim, J. 5, 10(15)
'limites naturelles' 40-2, 50,
55, 57-8, 69, 82, 135, 170;
see also frontiers
Lisbon 24, 27
Liska, G. 130(3), 137, 142,
154
Lithuania 46, 78, 82, 83, 88,
92, 97, 102, 103, T2, 121,
129, 131(14)
Livonia 80, 82
Lloyd George, D. 103, 130(1)
Loire, R. 38
Lorraine 49; see also Alsace-
Lorraine
Lotharingia 38, 61(8);
Lotharingian Axis 26,
29(17), 31, 35, 39, 44, 57,
58, 68, 126
Louis XIV of France 8, 39,
40, 41, 42, 43, 162
Low Countries 52, 55, 57; see
also Netherlands

Macedonia 13, 160-1
McEvedy, C. 28(2)
Mackinder, H.J. 9(3), 96-8,
101(37), 134-5, 141,
144(15), 169
macrocore 68, 69, 122-3, 125,
128-9, 170
Madrid 24, 149
Maghreb 13
Magyars, see Hungary
Mahan, A.T. 96, 101(35),
134, 135, 135-6,139
Malta, Siege of 18, 21
Mameluke Sultanate 17

Manchester, W. 51, 59,
62(27), 128
'Marca Hispanica' 20, 68
'Mare Nostro' 1(25), 60
'Mare Nostrum' 11(25), 170
marginal crescent (Mackinder)
96-7, 99
Maritime Crescent, see
European Maritime Crescent
maritime imperialism 5, 20-1,
56, 86
maritime power 18, 23, 26-7,
40, 42-3, 56, 59, 80-1, 83,
84, 85, 96-7; see also
'urge to the sea'
Maritsa, R. 14
Marlowe, Christopher 7,
10(22)
Marmara, Sea of 12, 13,
14-15, 16, 64, 93, 151
Marrioty, J.A.R. 62(22)
Marx, Karl 108-9, 122, 124;
Marxism 81, 102-3,
108-110, 112-3, 117, 119,
123-4, 125, 126-7, 128,
129, 133, 137, 139, 155-6,
157
Mecklenburg 62(26)
Mediterranean Sea 11(25), 13,
38, 68, 85, 103, 123, 125;
region 1, 5, 6, 10-11(24),
12-29, 38, 53, 94, 163
Meinig, D.W. 135
Melk 30
Mellor, R.E.H. 100(24)
Mersen, Treaty of (870) 61(8)
Meseta (Spain) 19, 23, 64,
98, 151
Mesopotamia 5, 13
Metternich, K.von 18, 61(6)
Meuse, R. 38
'mezhdurechie' 78, 86-7, 124
Middle East 12, 13, 16, 19,
140, 163
Middle Kingdom, see
Lotharingia
Milan, Duchy of 21
'Militärgrenze' 32, 33

Mitteleuropa 32, 35, 36, 41,
43, 50, 51, 55, 58-9, 60,
61(4), 105, 118, 121, 127,
140-1, 170
Mittelgebirge 47, 49, 50-1
Mittelmark 46
model of dominance, see
geopolitical model of
dominance
Modelski, G. 6, 10(16)
Mohacs, Battle of (1526) 13,
31
Mohammed II, Sultan (the
Conqueror) 13, 17
Moldavia T2
Mongols 13, 77-8, 82, 83, 87,
89, 90, 91, 98, 121, 124,
133-4, 141, 145(32)
Moors 24, 49; see also Islam
Morava, R. 13
Moravia 52
Moscow 43, 53, 76, 78, 80,
82-3, 84, 85, 86, 91, 104,
108, 111-12, 119, 121, 125,
125, 126, 149
Munich Agreement 52
Muscovy 76, 78, 80, 81-2, 83,
86, 87, 89, 91-2, 96, 98,
117, 119, 121, 122, 125,
145(32), 155, 161
Mussolini, Benito 8
'Musspreussen' 58, 170

Naples, Kingdom of 21
Napoleon I (Bonaparte) 8,
42-4, 47, 98, 148
Napoleon III 45
Napoleonic Wars 76, 80, 88;
see also Napoleon I
National Socialism (Nazism)
52-3, F3.4, 57-9, 148
nationalism 1-2, 6, 35-6, 39,
50, 52-3, 55, 57, 84, 86,
102, 103, 112-13, 119, 149,
151-2, 154-5, 162
nation-states 1-2; see also
nationalism
'Naturgrenzen' 50, 53, 58,
170; see also 'Limites
Naturelles'
Naumann, F. 62(29)

Nazism, see National Socialism
Nazi-Soviet Pact (1939) 53,
104
Netherlands 21, 25, 26, 32,
35, 40, 42, 49, 68
Neumark 46
'Neuordnung, Die' 59, 60,
63(46), 171
Neva, R. 77, 80, 85, 96
New Order, see Neuordnung
Newbigin, M.I. 16, 28(8),
Nicaea 12
Nicomedia 12
Niebuhr, R. 4, 8-9, 10(14),
119, 133
Nieman, R. 43, 46
Nizhni-Novgorod 82
Normandy 38
North European Plain 32, 46,
47, 49, 51, 68, 76, 103,
118, 121
North Sea 31, 38, 55, 58
Northern War 80, 97
Novgorod 77, 78, 82
Nystadt, Treaty of 80

October Revolution, see
Bolshevik Revolution
Oder, R. 46, 47, 49, 52, 65;
Oder-Neisse Line 105
Odessa 80
Oka, R. 78
Okhotsk, Sea of 126
Oldenburg 62(26)
Operation Barbarossa, see
Barbarossa
'oprichnina' 87
Orkhan Ghazi 12
Osman Ghazi 12; Osmanlis
12-13, 14, 16
Ostmark 30, 38, 68
ostrog 76, 171
O'Sullivan, P. 4, 10(11)
Otto I, Holy Roman Emperor
62(21)
Ottoman Empire 12-19, F2.1,
21, 23, 33, 55, 78, 83, 84,
91, 93, 98, 127, 145, 161;
development 12-13, 16-18,
151; dominant state
characteristics 64, T1, 68,

69, 70, 90, 98; extent
13-14, 136; fall 18-19, 21,
26, 27, 31-2, 85, 113, 148,
T4, 152; geographical
background 14-15, 18-19,
origins 12

Pan-Germanism 50, 55-6, 57,
58-9
Pannonian Plain 13, 18, 31-2,
33, 35, 61(6), 64, 118
Pan-Slavism 83, 84-5, 86, 92,
93, 123, 126, 143(9), 152
Pap, M. 130(2), 139-40
Paris 39-40, 44, 45, 52,
61(9), 65, 69, 151
Parker, G. 63(34), 132(37)
Parker, W.H. 101(31), 135,
144(16)(21)
Partitions of Poland, see
Poland
Partsch, J. 55, 62(30)
Patterson, J.H. 63(36)
Pax Romana 161; Russica 161;
Turcica 16, 161; see also
Imperial peace
perestroika 157, 159(11)
Peter I, Tsar (the Great) 80,
85, 87, 88, 157
Petrie, C. 29(29)
Petrograd, see Leningrad
Philip II of Spain 20, 24, 27,
28
Philip Augustus of France
38-9
Philotheus 99(11), 124
physical geography 5-6,
14-16, 17, 18-19, 25, 30-1,
32-3, 36-8, 43, 46, 47, 59,
76-7, 82-3, 87, 93, 148-9,
163
Piedmont 32, 42
Pivot theory (Mackinder)
96-8, 99, 171
Pleve, W. von 88
Pochlarn 30
Poland 43, 52, 78, 82, 88,
92, 102, 103-4, 105, 117,
118, 121, 129, 130(1), 149;
Partitions 32, 46-7, 49,
52, 80-81, 83, 104, 130(4)

Poliakov, L. 99(11)
Popes, see Catholicism
population 24, 35, 39, 43-4,
45, 47, 51, 56, 78, 87,
114-15, T2, 116
Portugal 20, 24, 29(18)
Potsdam Conference (1945)
123
Pounds, N.J.G. 62(15)(28),
99(8)
Preveza, Battle of (1538) 13,
18
Propontis, see Marmara, Sea
of
Provence 40
Prussia 36, 45-9, 50, 51, 52,
53, 56, 57, 58, 60, 65, 68,
69, 80, 81, 118, 142, 149,
161, 166; East 104-5
Prut, R. 121
Pyrenees 19, 23-4, 28, 40

racialism 9, 57, 59, 60, 149
Ramsey, W.M. 6, 7, 10(20)
Ratzel, F. 2, 10(6), 56
Reconquista 20, 21, 28
Reformation 21, 137
regnum 4
religion 4, 5, 8-9; see also
Christianity, Islam
Renaissance 6, 137
revolution, see Bolshevik
Revolution, French
Revolution
R.F.S.F.R., see Russia
Rhine, R. 25, 32, 38, 40-42,
44, 55, 57-8, 65, 151;
Rhinelands 21, 23, 40,
49-50, 51, 52, 55, 57-8,
68, 128, 150
Rhône, R. 38, 40, 58
Richelieu, Cardinal 41
Riga, Gulf of 52; Treaty of
(1920) 130(4)
Risorgimento 142
Robertson, C.G. 62(22)
Roman Empire 4-5, 8-9,
11(25), 17, 38, 41, 45, 60,
68, 94, 139, 141, 161
Romania 32, 33, 103, 105
Romanitas 41

Rome 5, 11, 17, 21, 25, 42, 84, 113, 124-5, 139; see also 'Third Rome'
Root, H. 28(7)
Rudolf of Habsburg 31
Ruhr 51, 58, 128, 148, 151
Rum, Sultanate of 12, 17
Rumeli 13, 14-15
Rurik 77
Russ, see Kievan Russ
Russia 32, 43, 47, 55, 56, 59; Russian Soviet Federated Socialist Republic (RFSFR) 112, 113, T2, 116, 119, 129, 154-5; see also Russian Empire, Soviet Union
Russian Empire 49, 52, 76-99, F5.1, F5.2, 103, F6.1, 149, 157, 161, 163; development 76-90; dominant state characteristics 89-94, F5.2, 97-9; fall 81, 88, 91-4, 98-9, 108; geographical background 76-7, 82-3, 86-7, 93; origins 76, 77-8; relation to Soviet Union 110-1, 117-9, 121-3; western views of 94-8
Russian Orthodox Church 78, 83-4, 86, 92, 123-4, 139, 155
Russo-Japanese War 98
Ruthenia 104, 105, 118

Saar 52
Sack, D.R. 4, 10(12)
St. Petersburg, see Leningrad
Sakaria, R. 12, 13
samoderzhets 139, 145-6(32)
Sarajevo 36
Sardinia 20
Sarkisyantz, E. 99-100(11), 125;
Scandinavia 52, 55, 125; see also Sweden
Schelde, R. 38
Schleswig 49
Schlieffen Plan 51

Schmidt-Hauer, C. 88, 100(28)
Schumpeter, J.A. 88, 92
Scythians 133, 142-3, 164
Sedan, Battle of (1870) 45
Seine, R. 38, 58, 65
Selim, Sultan 13
Seljuk Turks 12, 14, 17
Serbia 13, 31, 84
Seton-Watson, H. 100-1(30), 113, 130(12)
Sevastopol 80
Seven Years War 47
Seville 26, 27
Siberia 87, 98, 114, 116, 137, 140, 154
Sicily 20
Silesia 47, 49-50
Slavs 18, 30, 33, 35, 36, 46, 49, 51, 57, 60, 64, 77, 78, 83, 84-5, 86, 88, 89, 92, 94, 103, 119, 149, 152; see also Pan-Slavism
Solov'ev, V.S. 85, 100(15)
Smith, W.D. 62(32)
Sorel, A. 41, 62(14)
Soviet Central Asia 114-15, 116, 154; see also 'Evraziistvo'
Soviet Union (USSR) 52, 53, 58, 102-29, F6.1, F6.2, T2, F6.3, T3, F7.1, 152-8; Asian factor 94, 125-9, 133, 134, 137, 141, 142-3; comparison with Russian Empire 108-111, 117-8, 121-3; decline characteristics 152-8; development 104-8, 111-21; dominant state characteristics 118-29, 135, 139-42; economy 114, 116, 123, T3, 136-7; origin 102-4; population 112, T2, 114-6, 128, T3, 136; sphere of influence 105-8, 125, F6.2; views of 133-43
Spain, see Spanish Empire
Spanish Armada 26-7, 149

Spanish Empire 13, 18, 19-28, F2, 35, 36, 39, 43, 45, 76, 81, 142, 149; development 20-6; dominant state characteristics 64, T1, 68, 69, 70, 90, 98; fall 26-8, 40, 148, T4, 154; geographical background 19; origins 19-20
'Spanish Road' 21
Spate, O.H.K. 135, 144(17)
Spengler, O. 134
'Sprachboden' 57, 171
'Staatsidee' 2, 171
Stalin, Josef 108, 109, 112, 113, 124, 125, 128, 142, 154-5
Stalingrad 53; Battle of 104, 149
state, concept of 1-3
Stettin 47; Stettin-Trieste line 122, 127
Stockholm 85
Stoianovich, T. 6, 10(18), 88
Strasbourg 40, 41, 69
Struma, R. 14
Styria, Duchy of 31
Sudetenland 52
Suez Canal 55
Sulaiman I (the Magnificent) 13-14, 17, 19, 28-9(13)
Sully, Duc de 6, 27, 39, 162
Suzdal 78
Sweden 80, 83, 85, 87, 94, 97
Switzerland 8
Syria 13

Tagus, R. 19
Tajikstan 112, T2, 116
Tamburlaine 7, 133, 142
Tannenberg, Battle of (1422) 46
Tartars 12, 16, 77; see also Mongols
'Tausendjahrige Reich', see Third Reich
Teutonic Knights 46, 60, 82
Third Reich 52-3, F3,4, 56-60, 63(48), 104, 105, 118, 140-1, 148, 151, 152

'Third Rome' 84, 99-100(11), 113, 124-5, 143(9)
Tilsit, Treaty of (1807) 43
Timasheff, N.S. 134, 143(8), 143-4(10)
Timur Lenk, see Tamburlaine
Toledo 20, 23, 24
Tolstoy, L. 148
Toynbee, A. 9(3), 33, 44, 62(19), 124
Trafalgar, Battle of (1805) 43
Trajan, Emperor 5
Transcaucasia 81, 82, 112, 116, 119, 154
Transleithania 33-5, 36
Transylvania 32
Trend, J.B. 24, 29(24)
Trieste 122, 127
Trotsky, L. 103, 124
Truman Doctrine (1947) 166
'Tsargrad' 85, 93
Tullin 30
Turgenev, I. 93
Turkey 123, 152, 156; see also Anatolia, Ottoman Empire
Turkmen 112, T2 116

Ukraine 53, 84, 92, 97, 103-4, 112, T2, 116, 121-2, 153
United Nations Organisation 1
United States of America 53, 59, 60, 123, 125, 140, 155
Ural Mts 32, 59, 76, 77, 80, 87, 114, 116, 119, 121-2, 139, 153
'urge to the sea' 68-9, 80-81, 83, 84, 85, 89-90, 144-5(21)
U.S.A., see United States of America
U.S.S.R., see Soviet Union
Ustinov, P. 137-9, 145(28)
Uzbekistan 112, T2, 131(14)

Valdai Hills 77, 82, 87, 119
Varangians 133
Vardar, R. 13, 14
Venice 13, 18, 26

Verdun, Treaty of (843) 61(8)
Versailles 41; Treaty of 52, 59, 103, 123
Vichy France 53
Vidal de la Blache, P. 41
Vienna 13, 18, 28-9(13), 30, 31-2, 33, 35, 61(6), 69; congress of (1814-15) 42
Vikings 77
Vistula, R. 46, 47, 52
Vladimir 78
Volga, R. 53, 77, 78-80, 87, 91, 96
Volkhov, R 77, 83
Volk movement 56
'Volksboden' 57, 60, 171
Voltaire, F.A. 47
Vosges Mts 58
vozhd 139, 145-6(32)

Walters, R.E. 144(15)(18), 147
war 2-3, 7-8, 70-72, 87-8, 163-4; see also specific wars
Warsaw 47; Grand Duchy of 42
Warsaw Pact (1955) 108
Warthe, R. 46, 47
Wehrmacht 52-3, 59, 60, 104, 149, 166
Weigert, H.W. 134-5, 137, 144(12)(14)

Weimar Republic 52
Weltpolitik 56, 172
Weser, R. 49
West Germany, see Federal German Republic
White Russians, see Byelorussia
White, T.H. 131(20)
Whittlesey, D. 9(3), 139, 162
Wight, M. 4, 9(1)
Wilson, Woodrow 103
Winter War (1939-40) 104-5
Wittfogel, K.A. 23, 29(19)(22), 124, 151, 162
Wooldridge, S.W. 164
'World-Island' 1, 5, 9-10(3), 140
'World Process' 6
World view, see ideology
World War I 6, 51, 52, 55-6, 102-4, F6.1, 109, 117, 121, 123, 153; II 52-3, F3.4, 55-60, F6.2, 118, 121, 123, 125, 134, 153, 154-5

Yenisehir 12
Yalta Conference (1945) 105
Yergin, D. 131(28)
Yugoslavia 105-8, 123

Zollverein 49
Zone of Occupation 105

Printed and bound by CPI Group (UK) Ltd, Croydon, CR0 4YY

22/10/2024

01777620-0016